Identifying Future-Proof Science

Identifying Future-Proof Science

PETER VICKERS

OXFORD
UNIVERSITY PRESS

OXFORD
UNIVERSITY PRESS

Great Clarendon Street, Oxford, OX2 6DP,
United Kingdom

Oxford University Press is a department of the University of Oxford.
It furthers the University's objective of excellence in research, scholarship,
and education by publishing worldwide. Oxford is a registered trade mark of
Oxford University Press in the UK and in certain other countries

First Edition published in 2023

Published in the United States of America by Oxford University Press
198 Madison Avenue, New York, NY 10016, United States of America

British Library Cataloguing in Publication Data
Data available

Library of Congress Control Number: 2022935028

ISBN 978–0–19–286273–0

DOI: 10.1093/oso/9780192862730.001.0001

Printed and bound by
CPI Group (UK) Ltd, Croydon, CR0 4YY

For my parents

Contents

Preface

This book starts with something that looks more like philosophy of science, and ends with something that looks more like sociology of science. Or perhaps I should say integrated history, philosophy, and sociology of science. One reason is that the methods of 'pure' philosophy can be frustrating: they never seem to establish anything *definitively*. Debates seem destined to go around in circles, or else evolve somehow, without ever reaching a firm conclusion that might be held up to outsiders as a noteworthy achievement. I tried to add something important to the 'scientific realism debate' earlier in my career, fully imbued with the philosophy that *the truth is out there*, and the thought that just maybe I could help us reach that truth. But the tools at my disposal as a 'pure' philosopher never seemed to go very far. Whilst one could fill a career that way, I didn't want to just fill a career; I wanted to reach truth, or at least head clearly in that direction.

Thus I was drawn towards methods that were not merely philosophical. History seemed a good place to start, since with the history of science comes data, of a kind, that one might build a philosophy upon. Thus we reach 'HPS', a field premised on a thorough integration of history and philosophy of science. But what came to me much later was the thought that the methods of *sociology* might also be thrown into the mix. I had been averse to sociology, since the term 'sociology of science' always seemed to be attached to a specific (rather extreme) attitude towards science, as being so thoroughly influenced by social factors that there could never be any talk of 'facts', as I understood that word. But if we shake this specific movement off, and think more broadly about what sociology of science might be (social epistemology)—inspired by scholars such as Helen Longino—then another promising methodology presents itself. Just as HPS allows for a method that is partially empirical, so too sociology is no stranger to empirical methods. These methods bring the endeavour that bit closer to natural science, and move us that bit further away from 'pure' philosophy. In this way, one might still dream of saying something definitive about science, something that could draw a consensus of opinion in a way that is vanishingly rare in philosophy.

The extent to which I have managed to combine these methods, and say something rather definitive, is unclear. It remains predominantly a work in

HPS, and the contribution of sociological methods is meagre to say the least. But I hope that the reader will see value in the attempt, at least. And I hope that at least some readers will be inclined to pick up the baton and run with it. There are no doubt many holes in this work, as it stands. But if the fundamental methodology constitutes an improvement on 'the scientific realism debate', then it may save some readers years of toil, who would otherwise have adopted a methodology destined to lead them in circles, or at least nowhere definitive.

As for the holes, they remain despite my receiving an enormous amount of help along the way. Indeed, some scholars helped me to write this book to such an extent that if the culture of co-authorship were more like the natural sciences, then this monograph might have had twenty co-authors. I would first mention Kyle Stanford, who read the book carefully from beginning to end, and offered critical feedback weighty enough to reduce the number of chapters from ten to nine. Kerry McKenzie also read the whole thing, as did anonymous Reviewer A, each providing crucial comments (and crucial encouragement!). Mike Stuart and Hasok Chang both set up reading groups when I first had a full draft of the book, and each of these meetings brought immensely valuable feedback. I am hugely grateful to Mike, Hasok, and those in attendance at these two reading groups.

Some scholars carefully read one or two chapters. Juha Saatsi was a big help here, challenging me on Chapters 2 and 6, for example. Several scientists provided invaluable feedback concerning Chapters 5 and 7, including Gillian Foulger, Peter Schulte, Alessandro Chiarenza, Sean Gulick, Gerta Keller, Vincent Courtillot, Stephen Brusatte, and Sean McMahon. And concerning Chapter 7 specifically, I must thank the Geological Society of America (GSA), who assisted me in acquiring data about past GSA conferences. I benefitted from similar help courtesy of Jon Korman at the SVP (Society of Vertebrate Paleontology).

For help and advice on specific issues, sincere thanks (in no particular order) also go to: Teru Miyake, Naomi Oreskes, Neil Thomason, Thomas Rossetter, James Fraser, Robin Hendry, Nancy Cartwright, Wendy Parker, Joseph D. Martin, Timothy D. Lyons, Alexander Bird, Darrell Rowbottom, Henry Taylor, Douglas Allchin, Andy Hamilton, Ludwig Fahrbach, Maya Goldenberg, Karim Thebault, Omar El Mawas, Alex Broadbent, and Ian Kidd. A special vote of thanks to Manuel Galvão de Melo e Mota, who spent many hours providing me with rich information from the archives of the SPMicros/SPME (Portuguese Society of Electron Microscopy), far more information than I could ultimately use in the book, however fascinating.

I presented this material, in one form or another, at various venues over the years, and profited hugely from these experiences, both because I was forced to reframe the material for presentation, and also because I gained invaluable feedback from the audiences. Here I must give thanks to 3rd-year undergraduate students at the University of Durham, UK, who took Philosophical Issues in Contemporary Science between 2015 and 2021, as well as MA students who took Philosophical Issues in Science and Medicine. Thanks also to the Joseph Cowen Lifelong Learning Centre, as well as the 'Lit & Phil' (both in Newcastle, UK) where I presented relevant material in 2019. And thanks also to an audience at Johns Hopkins University, USA, where I presented relevant material, again in 2019.

Huge thanks of course to the funder, The British Academy, who trusted me with a Mid-Career Fellowship, which ultimately ran from 1 December 2019 through to 30 June 2021. And thanks to the administrative staff at the University of Durham research office, including Anna Hutchinson, Linda Morris, Eleanor Glenton, and Rachael Matthews, who worked hard to make the Fellowship a reality. This book simply wouldn't be here without that Fellowship.

Finally, most important of all was personal support, without which I could never have completed this project during the extraordinary stresses of the Covid-19 pandemic. Here I must mention the support of my parents, who have been solid rocks for me all the way through. I must also mention my running buddy in Durham, Chris Cowie—those runs were so important for mental health. But the last word must go to my extraordinary wife, and friend, Laura Vickers, who has been amazing in a thousand different ways, and who is, for me, a constant source of inspiration.

List of Figures

1

What Is Future-Proof Science?

1. Science and Scepticism

This book is about identifying scientific claims we can be confident will last forever. By 'forever' I mean so long as the human race continues, and assuming the scientific endeavour continues in a serious way, without some sort of apocalypse. For most purposes it is convenient to think ahead just 1000 years. A lot has happened in the development of human thought in the past 1000 years, needless to say. But I want to claim, and I want to argue, that some of our current ideas *will* still be with us in 1000 years, so long as the human race persists and that thing we call 'science' is not abolished by some well-meaning government body. This will strike many readers as hubristic, no doubt. It may well be asked, 'Who could dare to claim to know the minds of humankind 1000 years from now?' But a persuasive argument can be made, I believe, that *many* such scientific ideas can be identified, and so I hope to persuade many of those readers with a genuinely open mind, including those who start reading this book with a certain degree of scepticism. I agree that it is surprising—amazing, even—that we can rationally be confident that certain scientific ideas will remain intact 1000 years from now. Or even 5000 years from now. But in fact this is a reasonable thing to believe.

There are (at least) two very different reasons a scientific idea could last forever:

(i) We are stuck in a rut of human thinking out of which we will never escape. Our idea is totally wrong (or mostly wrong) but we are somehow prevented from seeing that, or even if we do see it we are unable to replace it with something better/truer.
(ii) Science has hit upon the truth, and all that remains is for scientists to build upon and develop the correct idea they already have. No feasible scientific developments could bring them to reject the idea.

It is the latter option, (ii), that I mean to refer to with the phrase 'future-proof science'. This isn't to say that (i) is impossible, and we'll take it quite

Identifying Future-Proof Science. Peter Vickers, Oxford University Press. © Peter Vickers 2023.
DOI: 10.1093/oso/9780192862730.003.0001

seriously in some later chapters. But what I mainly wish to argue is that some scientific ideas should be called 'facts', and they should be called 'facts' because they are true ideas—the universe really is the way the theory says it is (allowing for small adjustments). Moreover, we have overwhelming evidence for this, to such an extent that no feasible scientific developments could overturn it. For example, it couldn't ever be the case that we have the right idea, and lots of evidence, but somehow (by sheer bad luck perhaps?) we go on to accumulate lots of *contrary* evidence that is sufficient to overturn the correct idea we started with.

In short, this book argues that we have come to know things through science, beyond all reasonable doubt. Certain knowledge claims—the product of scientific labour—are justified, where by 'knowledge claims' I mean assertions of fact without any significant hedging or caveats. I hope even sceptics will grant that this is *possible*. Sometimes we can have knowledge where we didn't have it before. To give an example, we can come to know why the sky does not run out of rain. Further, it can be the case that we don't just have a *theory* about the rain, but that, over time, we have so much evidence for the 'water cycle theory' that it is not unreasonable to say that we are certain, and it is a fact. We stop talking about 'the water cycle theory', and simply talk about 'the water cycle'. If we meet a sceptic, it would not be *unreasonable* (though it may come across as patronising or arrogant) to say, 'I'm certain; I know that I'm right about this.' Of course, in social interactions it is often much preferred to 'agree to disagree', to respect somebody's opinions and beliefs. It is often much preferred to dial down one's confidence and say something like 'I think there's good evidence for this', as opposed to 'I know this is true'. But what may *seem* like objectionable hubris to your audience can sometimes be fully justified: it may be no exaggeration to say that you are sure (beyond reasonable doubt) that you are correct, and an alternative view is wrong, however uncomfortable it may feel to say this.[1]

I think it's worth expanding on this point about social discomfort a little further. In many cases we face difficult dilemmas vis-à-vis how we express our degree of confidence. For example, suppose you visit a music festival, and you're laid on the grass one evening staring up at the stars with a new friend. You hear them say, 'I guess we'll never know what those twinkly dots of light really are.' You might feel so awkward about contradicting your new friend, that you actually reply, 'Yeah, I guess not', even though (let's assume) you

[1] The concept of future-proof science is not inconsistent with 'epistemic humility'; see e.g. Kidd (2020) for a useful entry to the literature on humility and science.

studied astrophysics at university, and feel 100 per cent sure that scientists *do* know what stars are. The problem is, you just can't think of any way to contradict the person without coming across as patronising. It also doesn't really matter if you 'let it go' in this particular context.

In other contexts, this tendency to 'let it go' or 'agree to disagree' absolutely must be resisted. Sometimes it is crucially important to distinguish clearly between items of human knowledge, and issues that are unsettled and open for discussion, without hiding that distinction behind social niceties. If we swap the musical festival example for the Covid-19 pandemic, and we swap the statement for 'I guess we just can't know whether the AstraZeneca vaccine is safe', it becomes far more important to respond honestly instead of simply answering 'Yes, you might be right about that', or similar. Indeed, if you know a lot of about vaccine testing, it would be wrong *not* to challenge the statement; you might even end up saving the person's life. And in science generally there are plenty of high-stakes contexts where absolute honesty is paramount, and social niceties must be put to one side. To illustrate: scientists could not 'agree to disagree' with chlorofluorocarbon (CFC) companies in the 1980s on the question whether CFCs were causing ozone depletion. If scientists had agreed with the CFC companies that they couldn't really *prove* the link between CFCs and ozone, and didn't *really* know, and there was room for rational doubt, that would have been a death sentence—at the hand of skin cancer—for thousands of individuals who are alive today. A similar story can be told about the HIV-AIDS link (Godfrey-Smith 2021, pp. 311–12), and there were indeed many unnecessary deaths in this case—this isn't all merely hypothetical.

At the same time there is of course a sense in which we are never 100 per cent certain; a certain degree of doubt is always possible. Suppose I strike the keys of the laptop and say to myself, 'Do I really know I am typing right now? Do I really know that I am attempting to write the opening chapter of a book?' It's certainly *possible* that I am wrong. For example (as Descartes famously urged in the 17th century) I could be having the most vivid dream I've ever had. Or perhaps I am not asleep, but my senses—sight, sound, touch—are being manipulated in a way that is totally hidden from me (as in *The Matrix*). Or perhaps (back with Descartes again) even my thoughts are being manipulated, by some 'evil demon' or similar powerful being.

If we accept that these are (remote) possibilities, even for a case as rudimentary as whether I know that I am striking keys on my laptop, then it may be urged that I shouldn't say I am *sure*. I shouldn't say I am *certain*. At least not 100 per cent. And if not for everyday facts such as this, then *definitely* not for scientific ideas—such as the causal link between CFCs and ozone depletion—which are *much* further removed from everyday experience and the testimony of the senses. The problem with taking this line should be obvious however: if it is insisted that we aren't sure about scientific ideas on *these* grounds, then we have to accept that we are never sure about anything. In which case words such as 'fact', 'sure', and 'certain' are never applicable, and might as well be eradicated from the dictionary: 'Knowledge is impossible!'

In fact, those who urge scepticism about scientific ideas are usually absolutely clear that they are *not* 'radical' or 'global' sceptics. As Hoefer (2020, p. 24) writes,

> As philosophers of science we are entitled (and, I would say, obliged) to set aside radical skeptical doubts. Or to put it another way: once the scientific realist forces the anti-realist into positing radical skeptical scenarios in order to keep her anti-realist doubts alive, the game is over.

Thus scientific sceptics think it is reasonable to say that we know lots of things, especially everyday things such as that it is raining outside. Of course, we might be mistaken, and the drops on the window have come from the window cleaner. We might even be right, but for the wrong reason: it *is* raining outside, but the drops on the window that we used as evidence for our claim that it is raining outside actually came from the window cleaner—these are the 'Gettier' cases. But it is reasonable to say that we *know* when we have been sufficiently careful with our observations (e.g. we go outside and stand in the rain for five minutes). And this stands, even though it always remains remotely possible that we are asleep or are somehow being manipulated or otherwise deceived. As Van Fraassen (1980, p. 71) notes, 'we do in our daily life infer, or at least arrive at, conclusions that go beyond the evidence we have', and he is keen to hold on to such everyday conclusions: 'I must at least defend myself against this threatened [global] scepticism' (ibid.).

What sceptics wish to deny is that we can have a similar level of confidence in properly *scientific* ideas. Witness, for example, Brad Wray, who (clearly inspired by Van Fraassen) writes in his 2018 book *Resisting Scientific Realism*:

> I will argue that our current best [scientific] theories are quite likely going to be replaced in the future by theories that make significantly different ontological assumptions. (p. 1)
>
> I argue that there is reason to believe that many of our best theories are apt to be rendered obsolete in the future. (p. 2)
>
> We should not get too attached to our theories. (p. 65)

> Today's theories are as likely to be replaced in the future as were the successful theories of the past. (p. 65)
>
> [C]ontemporary scientists should expect that their scientific offspring will look back at their theories with the same attitude they have towards the theories of their predecessors. Their offspring [future scientists] will see that many of today's successful theories will have been discarded and replaced by new theories that today's scientists never even entertained accepting, theories that are currently unconceived. (p. 95)

These claims are purely concerned with science, and—just like Van Fraassen before him—Wray is clear (e.g. p. 43f. and p. 64) that he is *not* a 'radical' or 'global' sceptic. He has specific reasons for maintaining his scepticism about science whilst resisting scepticism in many contexts outside of science. Every scientific sceptic, or 'anti-realist', has to deal with this issue: where does their scepticism end? Under what circumstances, exactly, are they *not* sceptical? (See e.g. Stanford 2006, pp. 12–13.)

Naturally there is no absolute dividing line between scientific claims and other types of claim. It is not as if we reach scientific claims in one way—using the 'scientific method', say—and reach other claims in a completely different way. Wray and other scientific sceptics acknowledge that there is no clear dividing line, but this presents no problem for them: there can be a grey area and at the same time still be clear cases on either side. Sceptics argue that (many/most/all) claims on the scientific side are not secure, and we shouldn't make bold assertions about them (e.g. that they will still be in place in 1000 years). Claims on the other side of the divide may well be absolutely fine, and we might make bold assertions about *them*, even though it isn't totally impossible that we are dreaming, or our brain is wired up to a sophisticated alien computer.

By contrast, this book will argue that this is *not* the way to carve up what (not) to be sceptical about. The fact that an idea comes out of science definitely does *not* mean that we can't be just as sure about it as we can about many everyday things. The evidence for scientific claims can sometimes take a form quite unlike the evidence we have for more everyday claims, but that needn't block our ability to know things. Indeed, often scientific evidence can be *better*—for the purposes of making claims concerning what we know—than more 'everyday evidence'. Simply put, the scientific provenance of an idea has no bearing on how certain we can be about the future-proofness of that idea. Instead of looking at the provenance, we should look (directly, or perhaps indirectly) at the quantity and quality of evidence. And there are circumstances in which we can be sure that the evidence has crossed some threshold, such that it is no longer reasonable to remain sceptical about the underlying idea. There is no *exact* threshold, of course, and there will always be a time when the scientific community is split, with some (a significant percentage) willing to state that the evidence is in, and we should start using the word 'fact', and others (a significant percentage) insisting that we need to remain cautious about any such bold claims (see Chapter 7 for a contemporary example). But, sometimes, we get beyond that stage, and reach a time when at least 95 per cent of reasonable/relevant scientists are happy to use the word 'fact'. (The use of '95%' will be justified in due course.)

And, indeed, scientists sometimes want to make this point themselves. A highly respected National Academies Press publication contains the following:

> [M]any scientific explanations have been so thoroughly tested that they are very unlikely to change in substantial ways as new observations are made or new experiments are analyzed. These explanations are accepted by scientists as being true and factual descriptions of the natural world. The atomic structure of matter, the genetic basis of heredity, the circulation of blood, gravitation and planetary motion, and the process of biological evolution by natural selection are just a few examples of a very large number of scientific explanations that have been overwhelmingly substantiated.
>
> (Institute of Medicine 2008, p. 12)

But being able to list a few such examples is one thing; being comfortable with the crossover point where a claim becomes a fact is quite another. And this is no small matter. In the climate change literature scientists are constantly wrestling with this issue. One author of an IPCC Special Report recently asked, 'Where is the boundary between "established fact" and "very high confidence"?' (Janzwood 2020, p. 1668). For this scientist and many thousands of others, this book provides an answer.[2]

2. Misleading Evidence

Can scientific evidence be highly misleading? Can it be the case that the evidence *looks* extremely strong, to the extent that nearly all scientists want

[2] See also Hoyningen-Huene (2022), especially footnote 23 where he writes of Ernst Mayr, "Mayr often deplored that he was not aware that philosophers of science have investigated this transition from theory to fact."

to use the word 'fact', but that's only because the evidence has led them up the garden path? Certainly some have claimed this, citing examples from the history of science to support the claim. Alas, to my embarrassment, I have also said something far too close to this. In 2018 Stephen Harris at *The Conversation* got in touch with the philosophy of science group at Durham, looking for somebody to write an article on 'the biggest failed science projects'. This ultimately led to my article 'The Misleading Evidence that Fooled Scientists for Decades', published in June 2018 (Vickers 2018b), where I wrote 'history shows us that even very strong evidence can be misleading'.

This book will argue that, in the contemporary scientific world, evidence can never be all *that* misleading. At least, not if one is careful about it, as the scientific community always is in the fullness of time (so this book will argue). One of the primary examples in my 2018 article was something of a mistake, and I'll correct that mistake in Chapter 3 of this book. What I said in that article was not *totally* wrong(!)—it can be the case that one or two pieces of evidence can be very misleading, taken on their own, although even then the words 'fooled scientists for decades' are not warranted. Better would be 'fooled scientists temporarily', or 'fooled a few scientists, but not the whole scientific community'. The most obvious cases are those where an individual piece of evidence was very surprising, and perhaps had the *potential* to mislead the scientific community, but didn't. Crucially, scientists consider a whole body of evidence over a period of time; they are (usually) in no rush to make a knee-jerk reaction to an individual result. And it is vanishingly rare for a whole body of evidence to be misleading over a substantial period of time, at least in the contemporary scientific world, where there are so many scientists and so many different scientific teams ready to correct the mistakes, fallacies, unwarranted inferences, and exaggerations of one individual scientist or team of scientists. Thank goodness I did at least say, in the final paragraph of my 'Misleading Evidence' article, 'It's rare for evidence to be very misleading'. But this wasn't strong enough: a *whole body* of evidence is *never* 'very misleading' for a substantial period of time, and for a large enough, diverse enough, scientific community.

I have been talking about evidence as if it is one thing, but in fact 'evidence' is something of an umbrella term: evidence takes many different forms, in different contexts, and its quality and quantity can sometimes be very difficult to assess. I agree with Kyle Stanford (2011) when he writes that, 'Scientific confirmation is a heterogeneous and many-splendored thing; let us count ourselves lucky to find it – in all its genuine diversity – wherever and whenever we can' (p. 898). Evidential reasoning—in all its forms—cannot be represented by a single, simple equation, as the Bayesian model of confirmation would suggest. Much energy has been spent debating *empirical* evidence, most obviously evidence taking the form of accommodations and predictions of phenomena. But it is sensible, I submit, to use the word 'evidence' in a broader sense: we can have (good!) reasons for believing claims that are not straightforwardly empirical reasons. Evidence can sometimes take the form of an argument, for example. And evidence can sometimes come under headings such as 'consistency', 'coherence', and 'explanatory power': these are the so-called *non-empirical theoretical virtues* (see Schindler 2018 for a recent treatment). The intense focus (within academic literature) on successful *predictions* in recent decades is justified to a certain extent, since successful predictions can sometimes be very important individual pieces of evidence. But even several successful predictions can be overwhelmed by other considerations. How we weigh up all these different sources of evidence is far from obvious. Scientists on the ground often use their intuitions, and these intuitions are often quite reliable, though not always. My claim is not that we can come up with a formula for 'the weight of evidence' in a given case; far from it. My claim is merely that sometimes we are sure that the weight of evidence has crossed a threshold, and it is time to drop the word 'theory', and start using the word 'fact'.

When it comes to misleading evidence, it undoubtedly exists. But it exists just as much for everyday claims as scientific claims. Sherlock Holmes can be misled for a while, as all of the evidence seems to point to one guilty party, when in the end the culprit is somebody else. In fact, a huge number of books and films play on this kind of possibility. Very occasionally, evidence can be *highly* misleading in everyday life, as the world seems to conspire against us somehow. Rarely, somebody is out to deceive us, as Iago deceives Othello: Othello has good evidence that Desdemona is having an affair with Cassio, even though she is not. We can also imagine still greater deceptions which have nothing to do with science: e.g. how the producers deceive Truman Burbank in *The Truman Show*. In this case, Truman has extremely strong evidence for all kinds of things that are not actual—what he sees on the news is fictional, and all those around him know that it is, but act as if it isn't.

The senses can be thoroughly misled, too, even if they are incredibly reliable most of the time. I'm not talking about the way we seem to 'see' or 'feel' things in a dream—if that is misleading at all, it is an ephemeral deception, since we know it wasn't real as soon as we wake up. The senses can be misled more dramatically, for example when we fail to see the left-to-right lines in Figure 1.1 as parallel, horizontal lines. Or more dramatically still, we see the world very vividly as coloured, when it (almost certainly) isn't (Figure 1.2).

The colour illusion is particularly dramatic, because we can't reveal the illusion to ourselves as we can with the horizontal lines in Figure 1.1. Indeed,

Figure 1.1 A well-known optical illusion

Figure 1.2 Coloured strawberry in human experience of the world; uncoloured strawberry in the world

for thousands of years the human race was certain that the world is genuinely coloured, with only rare voices of (speculative) dissent. It was only with the rise of modern philosophy (the primary/secondary quality distinction), developments in physics (What are surfaces made of? What properties do they have?), and developments in psychology and neuroscience, that evidence gradually mounted that when it comes to colour, the world is not how it appears. So in fact, if one is looking for real cases of highly misleading evidence, for a whole community, over a long period of time, the best examples may come from *outside* science, and belong instead to the context where the scientific sceptics are not sceptical: everyday claims such as 'snow is white'.

As the book progresses we will look at various candidates for misleading evidence in the history of science. Numerous examples have now been put forward in the literature, cases where scientists were apparently fooled, and later had to change their minds. I will argue that such cases are *not* grounds for a strong form of scepticism, and leave open the possibility that we can identify many scientific ideas that are future-proof. Many contemporary scientific ideas will be excluded from this, of course, precisely because we have not crossed the evidence threshold yet (and we may never cross it). For one thing, even if the initial evidence looks good, it is prudent to reserve judgement until an idea has been rigorously tested. This has never been more obvious than with the recent 'replication crisis', where many results in psychology/medicine/social sciences, apparently based on statistically significant data, cannot be reliably replicated. The crisis shows clearly that sometimes judgements of the weight of evidence can initially be exaggerated, even by honest, professional scientists. But this is hardly evidence for the kind of scepticism this book is concerned with: it didn't take long for the scientific community to attempt replications of these studies, see those replications fail, and recognise that certain initial claims of 'strong evidence' had been exaggerated. The international scientific community wasn't for a moment tempted to form a consensus, or make an official knowledge claim, regarding these cases. Needless to say, examples of future-proof science identified in this book will be based on much stronger evidence than the cases at issue in the replication crisis.

3. Approximate Truth

Another important caveat before we really get started: I don't deny that there will be *adjustments* to scientific ideas in the future. Just about any scientific idea one can imagine will be subject to some kind of refinement over the next tens/hundreds of years. What I'm most concerned to resist, however, are claims that our current best scientific theories will be 'discarded' or 'rendered obsolete', as stated in the Wray (2018) quotations given above. Similarly, I'm keen to resist the claim, often made by the sceptic, that future scientists will take 'the same attitude' towards our current theories that *we* take towards past discarded theories, and that 'our own scientific theories are held to be as much subject to radical conceptual change as our past theories are seen to be' (Hesse 1976, p. 266).

To illustrate, consider models of our Solar System. One way to think about the history of such models is as follows. Ptolemy got it wrong: the Sun does not orbit the Earth—this idea was eventually discarded. Then Copernicus got it wrong: the Earth does *not* orbit the Sun in circular orbits. Then Kepler got it wrong: the Earth does not orbit the Sun in elliptical orbits. The latter idea is wrong, for example because the Earth's orbit is always perturbed by other bodies, such as Jupiter, but also because it assumes that the Sun's position is fixed, when it is not. Then the 19th-century Newtonians got it wrong, too: the Earth does *not* orbit the centre of gravity of the Earth–Sun system in a near-ellipse according to Newton's laws of motion. Einstein's general theory of relativity changed all that. And now it is widely assumed that Einstein's theory of general relativity needs to be quantised, somehow; this is what theories such as 'loop quantum gravity' are about. So we've been wrong wrong wrong. Each theory has been 'discarded', and along the way we've seen 'radical change' again and again, and we expect more.

Or have we? As I said, this is *one* way to think about the history of scientific thought vis-à-vis the Solar System. But it is contrived. Describing this sequence of theories in terms of repeated 'radical change' is misleading. Consider Newtonians such as Laplace, Poisson, and Le Verrier—specialising in celestial mechanics in the 18th and 19th centuries—faced with a philosopher of science saying,

> [C]ontemporary scientists should expect that their scientific offspring will look back at their theories with the same attitude they have towards the theories of their predecessors. (Wray 2018, p. 95)

Well, is this correct? *Were* those 19th-century Newtonian models of the Solar System just as subject to 'radical change' as the epicycle model of Ptolemy, including as it did a static Earth, with all other celestial bodies orbiting around it? Definitely not. Ptolemy's model of the Solar System was indeed radically false, in a large number of different ways—one cannot possibly shoehorn the term 'approximately true' onto this model. By contrast, the model Le Verrier was working with in the 19th century was exceedingly accurate. Contemporary scientists do *not* look back on Le Verrier's model with anything *like* 'the same attitude' that Le Verrier looked back on Ptolemy's model. And this is because—to put it bluntly—Le Verrier's model was approximately true. Absolutely no need for a shoehorn.

It may be objected: Le Verrier could never have dreamed that Einstein's theory of general relativity would come along, and completely transform our conceptions of space, time, and the meaning of 'gravity'. When it comes to space, time, and gravity, Le Verrier's views were indeed 'radically false', and eventually 'discarded' (at least as candidates for truth). But this is to shift the goalposts. We were talking about models of the Solar System, including what the Sun, the Moon, and the planets are, and how they relate to each other and interact with each other over time.[3] When I described Ptolemy's model as 'radically false', I was considering these respects, *not* his views on the nature of space and time. It goes without saying that there are always deeper questions one can ask, including, 'Is gravity a force?' But when it comes to modelling the Solar System one can choose to ignore such deeper 'metaphysical' questions, and get on with the modelling job, exactly as Le Verrier and many others did in the 19th century. And when one puts the deeper questions to one side and concentrates on assessing the model of the Solar System Le Verrier believed in, it cannot be denied that his model was approximately correct. In fact, many of the things he believed were *plain true*; for example:

> The Earth orbits the centre of gravity of the Earth–Sun system in a near-ellipse, subject to minor perturbations.

If one similarly looks for (significant, non-trivial) truths within the Ptolemaic account, one will struggle.

If we turn back to the concept of 'future-proof science', then, I do want this to be compatible with adjustments. Some of our current ideas will (of course) turn out not to be 'perfectly' true, but can reasonably be described as approximately true in the straight-forward way that Le Verrier's conception of the Solar System was obviously approximately true. No clever theory of 'approximate truth' is needed to substantiate this: I will use the term in the same way it is used in everyday life. We all handle the concept of *approximate truth* every single day of our lives, whether we realise it or not.[4] Different cases of

[3] How they interact *crudely speaking*, not at some deep metaphysical level. More on this 'depth of description' spectrum in due course.

[4] To illustrate: if we go out for dinner, and the waiter turns up to take our order and says, 'I'm ready to take your order' just as his hand is moving to his waistcoat pocket to retrieve his pad and pen, we will not object, 'Actually, you weren't ready when you said that. You're only ready now, some seconds later, when you've actually got your pad and pen in hand.'

application of the term 'approximately true' will come up in different contexts, as we progress, and as we tackle the case studies, so I won't say much more here (see e.g. Section 5 of Chapter 7). Suffice it to say, for now, that there are often clear cases of approximate truth in science, just as in everyday life. I submit that we will always look back on Le Verrier's model of the Solar System as an approximately true model. When I say that a scientific idea is future-proof, I do not mean that it won't change at all for the next 1000 years; I agree that there might be minor adjustments, just as there have been minor adjustments to some of Le Verrier's ideas about the Solar System. At the same time, however, some of Le Verrier's ideas are retained intact, and indeed this is always possible when the original ideas are approximately true. When I was looking for a statement from 19th-century celestial mechanics that was *plain true* I simply omitted reference to Newtonian mechanics. I also used the term 'near-ellipse', deliberately staying vague on how the orbit of the Earth varies from a true ellipse. Charles S. Peirce famously wrote, 'It is easy to be certain... One has only to be sufficiently vague.' What's crucial here is that one can often be just partially vague, still saying something of obvious substance. In this way it is often possible to be practically certain about something highly non-trivial.

4. Future-Proof Science

Which scientific ideas are future-proof? It is not my intention to use this book to provide a comprehensive list! But at the same time, I must be willing to step up to the plate and name some concrete examples. A good starting point is to provide some *singular facts* that are scientific in the sense that we know them to be facts as a result of scientific labour:

1. The Sun is a star.[5]
2. The Milky Way is a spiral galaxy, similar in structure to Messier 83 and NGC 6744.
3. The Earth is a slightly tilted, spinning, oblate spheroid.

[5] An anonymous reviewer asked 'What does this mean? How would you flesh it out?' (cf. the discussions in Fuller 2007, p. 10, and also Miller 2013, p. 1302). This same question could be asked of any one of my 30 examples. This issue will be addressed in Chapter 9, Section 2.4 ('Is the Sun a Star?'), but the brief answer is that one can use standard textbook definitions of key terms that are not super-detailed, but also far from trivial. It is worth reflecting briefly on the fact that 'Pluto is a planet' turned out *not* to be future-proof. However, Pluto was always an outlier, whereas 'our Sun is very much a run-of-the-mill star' (Noyes 1982, p. 7). Kinds and outliers will be further discussed in Section 2 of Chapter 8.

4. The Moon causes the tides (with just a bit of help from other factors, such as the pull of the Sun).
5. The collection of propositions summarised as 'The water cycle'.
6. DNA has a double helix structure.
7. Red blood cells carry oxygen around the body.
8. Normal person-to-person speech travels as a longitudinal compression wave through the particles in the air.

In these eight cases there can be no reasonable doubt. Indeed, these are such solid facts that any bona fide scientist—with relevant specialist knowledge—would find it absurd to add the word 'theory' to any one of these examples, e.g. to talk of the 'Water Cycle Theory'.[6] It may be objected that it is possible for an astronaut to directly *see* that the Earth is a spinning spheroid, but of course we knew the Earth was spherical long before that was possible (to the extent that it is). And in addition one can't say the same of *all* of these examples; we don't directly see that the Sun is a star.

If we think these are all indisputable facts, but direct observation doesn't provide the warrant, then why do we believe them so strongly? One answer is that we are taught that they are facts at school. But if pushed further we may agree that they are *taught* as facts because scientists have established that they *are* facts, over many decades, using a combination of scientific methods including observation, experiment, and theory-development. In short, the evidence for these eight claims has gradually built up until no reasonable doubt can be maintained. Very few of us actually know more than a *very* small fraction of the relevant evidence, and here an element of trust inevitably enters the picture. But—unless we are conspiracy theorists—we feel that this trust in authority is very highly motivated. (See Chapter 5 for a full discussion of the role of trust.)

If the given story is accepted, it is difficult to resist sliding a little further. If we accept what is taught to us at school as scientific fact—using that as a proxy for a huge amount of scientific evidence built up over many decades—then there are many possible examples, including more ambitious examples coming more obviously under the heading of 'scientific theory'. In fact, many such

[6] Cf. Hoefer (2020), p. 21: 'The core intuition behind SR [Scientific Realism] is a feeling that it is absolutely *crazy* to not believe in viruses, DNA, atoms, molecules, tectonic plates, etc.; and in the correctness of at least many of the things we say about them' (original emphasis). This book is *not* a defence of 'scientific realism', however; see Chapter 2.

examples were put forward in the philosophical literature in the 1960s and 1970s by those who wished to resist Kuhn's (1962) story of scientific revolutions, to make the point that his examples—exemplifying the cycle of 'normal science', 'crisis', and 'paradigm change'—were cherry-picked. As Godfrey-Smith (2003, p. 98) writes,

> [Kuhn] was surely *too* focused on the case of theoretical physics. [...] [I]f we look at other parts of science – at chemistry and molecular biology, for example – it is much more reasonable to see a continuing growth (with some hiccups) in knowledge about how the world really works. We see a steady growth in knowledge about the structures of sugars, fats, proteins, and other important molecules, for example. There is no evidence that *these* kinds of results will come to be replaced, as opposed to extended, as science moves along. This type of work does not concern the most basic features of the universe, but it is undoubtedly science. (original emphasis)

I couldn't agree more: a large part of our current understanding of sugars, fats, and proteins is surely future-proof, even if there remain many open questions about these molecules. And it is not only the structure of these molecules that we can claim knowledge of; we also understand a great deal about how they behave within the bodies of organisms, including human bodies. This is compatible with the thought that there remains much we do *not* understand.

Molecular biology is just the tip of the iceberg. Some scholars have countered the list of examples of rejected theories in the history of science with a list of examples of 'theories' or 'bodies of thought' that are apparently secure, and where no revolutions are even remotely anticipated. The following is a list of my own, building upon the eight examples already given (partly inspired by Fahrbach 2011, p. 152).[7] In each case I include a 'singular fact' that is illustrative of a wider body of claims coming under the relevant heading:

[7] Earlier scholars have also sometimes given their own examples of future-proof science (although they don't use that term). For example, McMullin (1984, pp. 27–8) gives examples from evolutionary history, geology, molecular chemistry, and cell biology. He also notes (p. 8) that, 'Scientists are likely to treat with incredulity the suggestion that constructs such as these [galaxies, genes, and molecules] are no more than convenient ways of organizing the data obtained from sophisticated instruments.' More recently, Hoefer (2020, p. 22) writes, 'There is a large swath of established scientific knowledge that we now possess which includes significant parts of microbiology, chemistry, electricity and electronics (understood as not fundamental), geology, natural history (the fact of evolution by natural selection and much coarse-grained knowledge of the history of living things on Earth), and so forth. It seems crazy to think that any of this lore could be *entirely mistaken, radically wrong* in the way that phlogiston theories and theories of the solid mechanical aether were wrong' (original emphasis). See also Hoefer (2020, p. 25f.) and Hoefer and Martí (2020).

9. Evolution by natural selection.[8]
    - Singular fact: Human beings evolved from apes that lived on Earth several million years ago.[9]
10. Numerous chemical facts about elements and how they relate to each other.[10]
    - Singular fact: A typical oxygen atom is 16 times heavier than a typical hydrogen atom.
11. The germ theory of disease, including numerous things we know about the properties and behaviour of various different bacteria and viruses, and how these sometimes contribute to disease and illness.
    - Singular fact: Syphilis is caused by the bacterium *Treponema pallidum* subspecies *pallidum*.
12. The 'neural net' theory of the brain, including a large body of knowledge vis-à-vis brain behaviour and the nervous system.
    - Singular fact: Visual input coming from the retina is processed at the rear of the brain.
13. Much of cosmology, including the large-scale structure of the universe, the expansion of the universe, and the properties of various entities such as quasars, pulsars, and galaxies.
    - Singular fact: Quasars were more common in the early universe.
14. A large body of thought concerning the geological history of our Earth, including (for example) knowledge of past ice ages.
    - Singular fact: Big Rock boulder in Alberta, Canada, was carried there from the Rocky Mountains by a glacier during the last ice age.
15. A large body of thought concerning the interior of the Earth, including knowledge of the inner and outer core.
    - Singular fact: The Earth has a liquid-metal outer core.
16. A large body of thought concerning the history of life on earth, including the 'Cambrian explosion', and the P-Tr and K-Pg extinction events.
    - Singular fact: There was an explosion of life on Earth approx. 540 million years ago.

[8] This will be tackled in Chapter 4. Of course, nobody would claim that natural selection is the *only* active mechanism.

[9] To get a sense of the state of the art, see e.g. Williams (2018); Böhme et al. (2019); and Almécija et al. (2021).

[10] The periodic table of elements is a tricky example in certain respects, since there are ongoing debates about how best to structure it (or at least, how best to structure *parts* of it); see e.g. Grochala (2018).

17. Detailed knowledge of the history of human life.
    - Singular fact: There have been several different human-like '*Homo*' species, of which only modern-day *Homo sapiens* remains.
18. Plate tectonics, including the history of past land-masses such as Laurasia and Gondwana.
    - Singular fact: Between 120 and 160 million years ago, South America split from Africa.
19. Knowledge of cells, mitochondria, chromosomes, and DNA.
    - Singular fact: The SRY gene on the Y chromosome is essential for the development of male gonads in humans.
20. Knowledge of the chemical and physical evolution of our Sun over the next six billion years.
    - Singular fact: Our Sun will gradually turn into a red giant over the course of the next six billion years.
21. Knowledge coming under the heading of 'biochemistry', including knowledge of the structure and behaviour (within organisms) of important molecules such as various sugars, fats, proteins, vitamins, caffeine, alcohol, etc.
    - Singular fact: Animal cells use glucose and oxygen to produce adenosine triphosphate, a high-energy molecule that can then provide muscles with energy to contract during exercise.
22. Knowledge of the structure of all kinds of molecules, and chemical reactions between molecules.
    - Singular fact: Vinegar ($C_2H_4O_2$) and baking soda ($NaHCO_3$) react to give sodium acetate ($NaC_2H_3O_2$) + water (H_2O) + carbon dioxide (CO_2).[11]
23. Detailed knowledge of many dinosaurs, including at least some aspects of how they lived and interacted.
    - Singular fact: Tyrannosaurus rex had a highly developed sense of smell.
24. Detailed knowledge of the properties and behaviour of sound waves.
    - Singular fact: Sounds waves are both longitudinal and transverse through solids, but only longitudinal through liquid and gas.

[11] As McMullin (1984, p. 28) notes, 'To give a realist construal to the molecular models of the chemist is not to imply that the nature of the constituent atoms and of the bonding between them is exhaustively known.'

25. Knowledge of the properties and behaviour of various different types of cancer.
    - Singular fact: Smoking causes cancer.
26. Knowledge of numerous illnesses and diseases, including Parkinson's, diabetes, epilepsy, HIV/AIDS, Huntingdon's, spina bifida, etc.
    - Singular fact: Human immunodeficiency viruses (HIV) kill immune system cells (T helper cells).
27. A large body of knowledge within pollen and spore science (palynology).
    - Singular fact: Endospores can stay dormant for millions of years.
28. Thermodynamics.
    - Singular fact: At a constant temperature, the pressure of a gas is inversely proportional to its volume.
29. Numerous facts coming under the broad heading of 'climate science', including human-caused global warming.
    - Singular fact: The concentration of carbon dioxide in the Earth's atmosphere in the year 2020 was the highest it has been in three million years.
30. Materials science: our understanding of properties and behaviours of various different metals, alloys, plastics, etc., going far beyond purely empirical knowledge.
    - Singular fact: Polycarbonate molecules absorb UV radiation.

So, I think it is quite easy to give 30 examples,[12] even including some very broad examples which actually include within them numerous more-specific scientific facts/theories. Of *his* list of nine examples, Fahrbach writes: 'Despite the very strong rise in amount of scientific work, refutations among them ["our best scientific theories"] have basically not occurred' (p. 151). The significance of the 'very strong rise in the amount of scientific work' will be explored in Chapter 2, Chapter 5, and elsewhere.

Of course, the sceptic will absolutely expect to see a (long) list of 'current best theories' that have not (yet) been refuted. It is hardly evidence for

[12] There is some overlap in my examples; e.g. examples 2 and 13, and examples 8 and 24. It is no struggle to come up with additional examples, however. For example, I haven't included Hoefer's (2020) examples concerning (i) our knowledge of electrical phenomena (at a non-fundamental level of description), and (ii) nuclear physics, including facts about nuclear fusion and fission, and nuclear (in)stability. Throughout this book I will repeatedly refer to 'the 30 examples from Chapter 1', with the thought that any examples that concern the reader could easily be replaced with alternative examples.

future-proof science that one can produce a long list of current theories concerning which current scientists are confident. Lord Kelvin, at the turn of the 20th century, reportedly stated that, 'There is nothing new to be discovered in physics now. All that remains is more and more precise measurement.' And Albert A. Michelson (famed for the Michelson–Morley experiment of 1887) wrote in 1903:

> The more important fundamental laws and facts of physical science have all been discovered, and these are so firmly established that the possibility of their ever being supplanted in consequence of new discoveries is exceedingly remote. (Michelson 1903, p. 23f.)

Given that Kelvin and Michelson said these things, their own lists of examples of 'future-proof science' would no doubt have included examples of 'classical' 19th-century physics that we have now quite thoroughly rejected (at least as candidates for truth). So, we have to be careful: the fact that some prominent scientists are confident about an idea, or theory, should not *by itself* convince us that the idea is (probably) future-proof. But that's OK: this isn't the reason I am confident about the 30 examples listed above. The reason I am confident has to do with the quantity and the quality of the evidence for these ideas, vetted by thousands of scientists, embedded within a sufficiently diverse scientific community.

That's the (very) short story. The long story is rather more complicated, and will be filled in gradually over the next eight chapters.

5. Outline of the Book

It is time to get stuck into the details of the debate. This we turn to next, in Chapter 2. So far I have only sketched the position of the 'scientific sceptic', and there are importantly different sceptical positions. Indeed, some of the scholars who describe themselves as 'sceptics', or 'anti-realists', or 'instrumentalists', actually hold positions extremely close to my own. This sounds backward, but that is only because of a confusing use of labels in the relevant literature. It is also crucial for me to engage with the so-called 'scientific realism debate'. I actually do *not* consider this book a stance in the scientific realism debate, since that is a debate most usually defined by a particular distinction between 'observables' and 'unobservables', which will not matter much here, and which I believe to be unfortunate. At the same time, I do wish to argue against the proclamations of many 'anti-realists' or 'non-realists' (including Wray, Stanford, and Van Fraassen).

Following the philosophical groundwork of Chapter 2 we move on to various case studies from both the history of science and also contemporary science. Chapter 3 is the first of the historical case studies. It concerns JF Meckel's (1811–27) novel predictive success concerning the existence of gill slits in the mammalian (including human) embryo. It is argued that this successful prediction, whilst prima facie impressive, only modestly confirmed Meckel's theory of recapitulation. This demonstrates that there is no clear link between novel predictive success and truth, even if novel predictive success can sometimes be extremely influential as a type of first-order evidence.

Chapter 4 continues the story of novel predictive success as a candidate example of highly persuasive first-order evidence. Whilst Chapter 3 shows that novel predictive success cannot always be relied upon as a hallmark of future-proof science, Chapter 4 argues further that novel predictive success can be rather *insignificant* evidentially speaking, even when it appears very significant. It does this via a discussion of a relatively recent novel predictive success of the theory of evolution, one that has been selected by contemporary scientists as a significant piece of evidence for the theory: the 2004 discovery of the 'missing link' fossil *Tiktaalik*. Chapter 4 argues that it is much better to direct attention away from individual successes such as this, and towards the full body of evidence. Whilst the full body of evidence is in practice inaccessible, even to senior experts in the field, it is argued that the weight of evidence can be judged *in*directly via a consideration of certain features of the relevant scientific community. This marks a turning point in the book, with future-proof science being identified via second-order, not first-order, evidence.

If we really turn away from first-order scientific evidence we must ask ourselves afresh: why do we firmly believe various scientific claims, such as the 30 examples listed in the previous section? The answer seems to be that we trust in scientific community opinion. Thus in Chapter 5 we start to ask the question: under what circumstances is scientific community opinion a hallmark of future-proof science? This leads to another historical case study, this time concerning a case where scientific community opinion apparently *got it wrong*: the case of continental drift 1915–65. It was supposedly proven impossible for the continents to move; many scientists believed this result, and thus continental drift research was ridiculed and otherwise inhibited or suppressed. Does this mean that scientific community opinion can*not* be confidently linked to future-proof science? Chapter 5 analyses the continental drift case and argues that it *can* be so linked, but we need to carefully identify sufficiently strong cases of scientific consensus. Put briefly, I require a solid scientific consensus amounting to at least 95 per cent, in a scientific community that is large, international, and diverse.

Chapter 6 addresses Hoefer's (2020) concern that, when it comes to fundamental physics, there is a 'special vulnerability to underdetermination', demanding significantly greater epistemic caution compared with other scientific contexts. Indeed, Hoefer's argument would suggest that, when it comes to 'future-proof science', one ought to treat fundamental physics as a very special case, completely blocking all pertinent claims, not because they are *not* future-proof, but because one can't be sure. Chapter 6 starts by demonstrating the problem via a discussion of Sommerfeld's 1916 prediction of the hydrogen fine-structure spectral lines, based on a radically false theory of the atom.

It is agreed that there are special epistemic problems in this context, but Hoefer's particular way of drawing the distinction—contrasting 'physics' and 'fundamental physics'—is shown to be problematic: for one thing, the concept *fundamental* can't bear the weight Hoefer wishes to place upon it. Alternative options are considered, including Van Fraassen's (1980) observable/unobservable distinction. But in the end it is argued that any such epistemic distinction will always be too crude, too sweeping. Instead we do better to trust the relevant scientific community—who are already highly cautious in this context—to decide on a case by case basis. Thus it is argued that the criteria for future-proof science introduced in Chapter 5 are also reliable in the context of 'fundamental physics' (broadly construed), and no special caveat is needed.

At this point in the book the link between scientific community opinion and future-proof science has been argued. But there are holes yet to fill in, and these come to the fore when we attempt to apply the proffered theory of future-proof science to contemporary cases. In Chapter 7 we turn to one of the most intriguing hypotheses of recent decades: the asteroid impact theory of the extinction of the dinosaurs. Many scientists have been tempted to state the hypothesis as a fact, and in 2010 a review article was published in *Science* hinting at a scientific consensus. There was a significant community reaction against this piece, however. In addition, there has been plenty of opposition to the claim in both the published literature and activity at (some) major conferences, all the way through from 1980 to 2020. This chapter navigates some of the challenges that can arise when we ask after the strength of feeling in the relevant scientific community vis-à-vis a specific claim. The case carries important lessons for how scientists go about declaring a consensus of opinion, a matter of crucial importance if—as this book argues—we are to identify future-proof science via sufficiently strong scientific consensus.

Chapter 8 applies the proffered theory of future-proof science to another contemporary case, this time of great social importance. During the Covid-19 pandemic, billions of people urgently wanted, and needed, answers to questions concerning scientific knowledge. Were all of the deaths definitely linked via a viral cause? Did it definitely originate in China in December 2019? Were the vast majority of children really safe? Could the vaccines be trusted? One thing lacking was a clear account of how the individual (whether expert or non-expert) could identify the future-proof scientific claims (the 'facts'), distinguishing them from other types of scientific claim, such as 'promising hypotheses', or 'useful speculations'. Looking to the criteria for future-proof science put forward in this book, a worry arises that nothing scientists were saying, in 2020, about the pandemic, could responsibly be called 'future-proof', since in 2020 so little time had passed for relevant scientific claims to be internationally scrutinised. But scientists did in fact have some relevant future-proof knowledge, even only a handful of *weeks* after the onset of the pandemic. This chapter explains how this is possible, given that usually absolute confidence in scientific claims depends upon extensive international scrutiny, often taking many years.

Chapter 9 articulates my final proposal for identifying future-proof science. It draws on the lessons from all the previous chapters to lay out (i) the criteria for future-proof science, (ii) the core argument supporting these criteria, and (iii) a workable strategy for actually *identifying* future-proof science. I build on the 'externalist' suggestion put forward by Oreskes (2019) that the best strategy is to use certain tools to critically assess the status of the scientific consensus, as a proxy for evaluating the entire wealth of first-order evidence from a large number of different perspectives. The shift from 'internal' evidence to 'external' evidence supports calls for adjustments to science education in our schools, with greater emphasis on teaching the 'external', second-order, or 'sociological' evidence for scientific claims. Additionally, this chapter raises some possible, outstanding objections, and provides preliminary responses.

2

The Historical Challenge to Future-Proof Science

The Debate So Far

1. Frustration and Miscommunication in the 'Scientific Realism Debate'

'Antirealism' is a position defined by its opposition to realism. And realism, on the face of it, is the position according to which we have good grounds for believing that the entities and processes posited by our best scientific theories are *real*. So the antirealists by contrast claim that we do *not* have good grounds for believing that the entities and processes posited by our best scientific theories are real. Incredibly, this is *already* a misleading construal of the debate. Over the past 10 years I have been repeatedly astonished at how easy it is to misinterpret what is happening in the realism/antirealism debate. Even professional scholars are regularly talking past one another. God help the non-specialist, who is just dipping into this debate from the sidelines. And in that I include scientists themselves; on the rare occasions that scientists take the trouble to try and see what is happening in the realism/antirealism debate, they will no doubt find it all very confusing, or misguided, and soon find themselves saying 'Nothing for me here'.

The problem can be nicely illustrated by a real-life encounter between practising scientists and 'the realism debate' in 2015. Consider first that 'antirealists' or 'nonrealists' often say things such as:

> We must really guard ourselves against believing forever warranted those hypotheses which have become universally adopted conventions [. . .]. The history of physics shows us that very often the human mind has been led to overthrow such principles completely, though they have been regarded by common consent for centuries as inviolable axioms.
>
> (Duhem 1906, p. 212)

Identifying Future-Proof Science. Peter Vickers, Oxford University Press. © Peter Vickers 2023.
DOI: 10.1093/oso/9780192862730.003.0002

> Our own scientific theories are held to be as much subject to radical conceptual change as our past theories are seen to be.
> (Hesse 1976, p. 266)

> [O]ur own historical successors will someday view even the leading scientific theories of our own day in very much the same way that we regard those of our historical predecessors. (Stanford 2015a, p. 412)

> [O]ur theoretical conceptions of nature will continue to change just as profoundly and fundamentally as they have in the past.
> (Stanford 2015b, p. 875)

> [T]here is reason to believe that many of our best theories are apt to be rendered obsolete in the future. (Wray 2018, p. 2)

These quotes—spanning more than 100 years—appear to say quite bluntly that we ought to consider our current best scientific theories as merely pragmatically useful, convenient fictions, or instruments for manipulating the world around us and (perhaps, in some cases) constructing psychologically satisfying how-possibly explanations. We should *not* take them to be literally true 'future-proof' accounts of how the world really is, providing how-actually explanations, and we should expect that they will probably be rejected one day (or at least *many* or *most* of them will). It is then quite natural to ask the question how practising scientists feel about the matter. What proportion of practising scientists would agree with this assessment?

In 2015 this was put to the test: a PhD student from the LSE, working on issues broadly connected with realism/antirealism, but coming from a geology background, decided to canvass the opinions of practising geologists via a questionnaire. Question 4 ran as follows:

> Because we cannot directly observe entities and processes in the geological past, some philosophers of science contend that they cannot be said to exist in reality. Do you think they are correct?

And the answer? Only four out of 93 answered 'yes'. Out of the other 89, 'some were affronted by the suggestion that what they were studying might not be real in some way'.[1] In other words, they very strongly disagreed.

Scientists will quite rightly be shocked by the thought that if something cannot be directly observed we shouldn't say that it exists in reality. If we

[1] Personal communication with the student behind the questionnaire.

should be sceptical of anything that cannot be directly observed, what's to stop us applying that to the Holocaust? Obviously, the evidence that the Holocaust really happened is so overwhelming that it is gut-wrenchingly abhorrent to suggest that there is room for doubt. It is technically true that we can't be 100 per cent certain: it's not as if we can 'prove it like a theorem'. But that is purely academic: sometimes the direct/indirect evidence for something being an element of reality—either right now, or in the past—is so overwhelming that it is beyond all reasonable doubt. And then we should call it an item of human *knowledge*, or a *fact*. Accessing that evidence is easier for some than others, but clearly historians/philosophers of science are well placed to access relevant evidence. So one can absolutely expect practising geologists to feel affronted when faced with the idea that some historians/philosophers of science want to say that tectonic plates, the core, mantle, and crust, and so on, are probably *not real things*, but are instead 'nice ideas' current scientists have that will most likely one day be rejected. They would be even more affronted if told that they only believe these are 'facts' because they don't know the history of science like we do, as if historians and philosophers of science are authorities who are in a position to educate the geologists on the reality of the entities and processes those geologists study. Consider Oreskes (2019):

> [D]espite the claims of prominent scientists to the contrary, the contributions of science cannot be viewed as permanent. The empirical evidence gleaned from the history of science shows that scientific truths are perishable. [...] Weinberg is a brilliant man. [...] But this comment reflects either a shocking ignorance of the history of science, or a shocking disregard of evidence compiled from another field. (p. 50, and fn. 88)

The 'other field' she is referring to is History. Oreskes is well known for *defending* science against scepticism, of course (more on this in due course).

This state of affairs is embarrassing for the (history and) philosophy of science community and prompts the question: has something gone wrong? The truth is, *hardly any* philosophers describing themselves as 'antirealists' or 'instrumentalists' are sceptical about the broad-brush/crudely stated claims of many of our current best geological theories. These 'sceptics' *really do believe* in the reality of tectonic plates and continental drift, and they *really do believe* that the inner core of the Earth is solid metal, and the outer core is liquid metal (though of course there are many open questions regarding the finer details). These are just a couple of examples of a great many contemporary geological theories that most 'antirealists' believe to be (approximately) true, and that they absolutely believe are *not* 'as much subject to radical conceptual change as our past theories' (Hesse), or 'apt to be rendered obsolete in the future' (Wray).

So what has gone wrong? The PhD student who designed the survey shouldn't be criticised here; the Hesse, Stanford, and Wray quotes (above) are typical of what one finds in much of the antirealist literature, and such quotes really do suggest that scientific antirealists advocate scepticism vis-à-vis unobservable entities and processes, such as those within the present/past Earth studied by geologists. Indeed, Wray's whole book *Resisting Scientific Realism* (Wray 2018) is explicitly against the idea that 'we have adequate grounds for believing that our theories are true or approximately true with respect to what they say about unobservable entities and processes' (p. 1). Why *wouldn't* one imagine that the entities and processes of contemporary geological theory count as such 'unobservable entities and processes'?

Most antirealists will probably say that I have quoted them out of context. They will say that if anyone reads their *whole* book/article, they will see that antirealist scepticism does not apply to such geological theories. And it does not apply (many of them would say) because these theories are concerned with *observables*.[2] The tectonic plates are observ*able* (even if not observed), because they are huge macroscopic bodies that in principle (if not in practice, except very indirectly) *could* be observed. And so too the inner and outer core of the Earth, including their dimensions, and their solid/liquid/metallic properties, *could* in principle be observed, even if they never will be in practice (except very indirectly). Thus such geological theories, being concerned as they are with observables, are suitable candidates for belief, and knowledge claims, once sufficient evidence has accumulated. For example Wray—pointing to Van Fraassen (1980) for anyone interested in the details—writes simply that antirealists are sceptical about 'the claims our theories make about unobservable entities and processes' (2018, p. 49), and anyone who's read Van Fraassen will know that 'unobservable' means 'unobservable in principle' (not in practice).

This could be OK. Says the antirealist: 'Perhaps some of my quotes are misleading taken on their own, but they should *not* be taken on their own.' However, this is not an adequate antirealist defence, for at least two reasons.

First (i) antirealists often leave large gaps between the statements they make concerning the limits of their scepticism in terms of in-principle unobservables (or whatever), and the statements they make about our current

[2] Kyle Stanford is an exception; I'll discuss his views shortly.

best theories being 'rendered obsolete' in the future, replaced with radically different theories that are currently unconceived. This causes frustration in the community since the statements about our current best theories being 'rendered obsolete' are widely seen as overly dramatic exaggerations when they do not include the caveats concerning in-principle unobservables, along with other obvious qualifications such as 'many of' and 'probably'. This encourages scholars to talk past one another, and obstructs genuine progress in the debate.

Second (ii) antirealists often are not clear what precisely is meant by 'unobservable'. If one digs into it (e.g. one reads Wray, gets directed to Van Fraassen, and finds the relevant discussions in the literature, e.g. Churchland and Hooker 1985), one finds that the word 'observable' is used very broadly indeed. For example, Van Fraassen has no trouble believing that dinosaurs existed, and that we know lots of things about them—dinosaurs are 'observable', because they are the kind of thing human beings are *in-principle* capable of observing. So too when it comes to tectonic plates and the Earth's core, or planets orbiting distant stars, or even the evolution of *Homo sapiens* from fish over many millions of years. Of course, many of those outside the debate would baulk at the suggestion that the evolution of *Homo sapiens* from fish is 'observable', but 'observable' is a technical term in the debate, not a natural language term. That's OK, but it is apt to cause confusion; for example, Wray (2018) relies on the observable/unobservable distinction, but doesn't discuss it anywhere in his book, preferring instead to reference Van Fraassen (p. 49) and Stanford (p. 100).

When antirealists do define 'observable', it can still sound as if they would *not* believe any/many contemporary geological theories. Wray (2018, p. 58) equates 'theoretical knowledge' with 'knowledge of unobservable entities and processes', and in a discussion of Hesse (on p. 64) he states simply that theories consist of 'theoretical claims', and it is these 'theoretical claims' which are 'most likely false'. He adds later (on p. 85) that what he rejects is 'the realists' claims about the growth of *theoretical knowledge*' (emphasis in original). If we wonder what Wray is *not* a sceptic about, the answer (on pp. 64–5) seems to be 'observation sentences' (following Hesse 1976, p. 274). Hesse writes:

> [T]here is accumulation of true observation sentences in the pragmatic sense that we have better learned to find our way about in the natural environment, and have a greater degree of predictive control over it [. . .] [T]his formulation of the growth of science does not presuppose privilege for our theory, because it is consistent with replacement of whole conceptual frameworks, including basic classifications and property assignments. (p. 274)

But let us now apply this distinction between 'theory' and 'observation statements' to the inner and outer core of the Earth. Our knowledge here is acquired primarily via analysis of seismic waves and the Earth's magnetic field. Any observations we make are observations of wave properties and magnetic field magnitudes. So if we take Hesse's words at face value, it seems clear that she would *not* agree that we have genuine knowledge of the properties of the inner and outer core; after all, none of our 'observation statements' concern the properties of the inner and outer core. It seems right to say that our current best models of the nature and behaviour of the inner and outer core are theories based on evidence, where that evidence consists in the behaviour of the seismic waves and magnetic fields we can directly measure. If this is right, then Hesse's scepticism *does* extend to the properties of the inner and outer core—she would *not* say boldly that we know the Earth has a solid inner core and a liquid outer core. She would instead say that these are just pragmatically useful theoretical ideas, allowing us 'to find our way about in the natural environment, and have a greater degree of predictive control over it'. And Wray (2018, pp. 64–5) quotes her with approbation.[3]

The word 'theory'—just like 'observable'—is very tricky in this literature. Apparently Hesse and Wray wish to use it to refer to scientific claims that concern unobservables: for them, if a claim is theoretical, then that *means* it concerns unobservables and is thus subject to scepticism (*whatever* evidence comes in). A more natural way to use the word 'theory' is simply to mean that one is sceptical that *enough* evidence has come in, so far, to allow us to use words like 'fact' or 'knowledge'. Turning back to our inner/outer core example, there was a time in the 20th century when it was perfectly natural for scientists to say: 'The claim that the outer core is liquid metal is currently just a theory; we need more evidence before we can be sure about it.' But on *this* conception of 'theory', there came a point during the 20th century when all reasonable geologists were happy to say, 'We know this now; it isn't *just a theory* anymore.' There is ample room for confusion here, since both uses of the word 'theory' concern reasons to be sceptical: on the one hand, because of claims about *unobservables*, and on the other hand, because of a lack of evidence. The crucial difference, however, is that on the former construal (but not the latter) the scepticism is there to stay.

Most antirealists, I submit, are open to believing in scientific theories that concern 'observables', broadly construed. These 'antirealists' might even

[3] See Chapter 6 of Bird (2022), 'Observation', for a similar discussion.

consider all of the 30 examples provided in Chapter 1 as concerned with 'observables'. In which case, all of the examples are *irrelevant for the scientific realism debate*: the antirealist may be just as willing to believe in them as the realist. To really stress the point, it is possible that the self-described 'anti-realist' is *more* willing to believe that those 30 examples are future-proof than the realist! When it comes to cases such as this, concerned as they are with 'observables' (broadly construed), it is simply a matter *for both the realist and the antirealist* whether the evidence is sufficient for belief. And it could be the case that an antirealist thinks the evidence is sufficient, for whatever reason, whereas a realist does not. A realist can be very cautious concerning the amount of evidence required before a knowledge claim should be made—why not? It is hardly a defining feature of the realist position that knowledge claims are made incautiously.

These few comments are illustrative of the way in which the realism debate has been full of frustration and miscommunication for several decades now (cf. Stein 1989). The observable/unobservable distinction has been instrumental in presenting the debate as one between protagonists who are widely opposed on the relationship between science and truth, when in fact there is much agreement. Indeed, scientific scepticism is usually significantly circumscribed. The more it is circumscribed, the closer together 'realists' and 'anti-realists' become.[4]

2. Stanford's Scientific Scepticism: Death by a Thousand Qualifications?

Kyle Stanford's book *Exceeding Our Grasp* suggests quite strongly that our knowledge claims typically go too far, exceeding what we can reasonably claim to know. The primary argument of the book concerns a historical lesson: there have been genuine cases in the history of science where one particular scientific theory seemed to have 'won the day', but it later turned out there was an *unconceived alternative* that was ultimately much better, and would come to reign supreme in that same domain for a later scientific community. There is an obvious implication here: perhaps, although our current scientific theories have 'won the day', they have only won it *for now*, and there is an unconceived

[4] There is a long and complicated story to be told concerning how the realism debate has evolved over the decades to end up in the state it is now in. For an entry to this historical story, see Saatsi (ed.) (2018), Part I: 'Historical development of the realist stance', and Vickers (2019) for my own take on it.

alternative that a later scientific community will adopt, rejecting as they do so our current theory. In other words, if we look at the pattern of scientific change over the history of science, there is reason to be extremely cautious about predicting the future. We should expect our current theories to be ultimately rejected, cast out, and perhaps even 'rendered obsolete'. As we read in Stanford's book right at the beginning, on p. 3, even before we start reading Chapter 1, 'Theories come and theories go. The frog remains.'[5] This clearly suggests to the reader that we should expect even the theory of evolution to 'go'.

Although a *quick* read-through of *Exceeding Our Grasp* suggests the strong form of scientific scepticism just sketched, a rather *careful* read-through does not.[6] Far from it, in fact: Stanford's scepticism (just like Van Fraassen's) is very much circumscribed, most often applying to entities and processes that *any cautious realist would also be inclined to be sceptical about.* In other words, Stanford and many 'realists' are in agreement concerning both (i) many scientific results we *should* believe, given the evidence, and also (ii) many scientific results we *should not* believe, given the *lack* of evidence.

This time the scepticism is circumscribed in a different way—Stanford does not wish to put weight on the observable/unobservable distinction (Stanford 2006, p. 34). He writes, 'the historical challenge is correctly focused not on our beliefs about entities of a particular sort (i.e., "unobservables"), but instead on beliefs arrived at or justified in a particular way' (p. 35). The particular 'way' he has in mind is an 'eliminative inferential procedure' (p. 28ff.), where 'we arrive at a decision to accept or believe a given theory because we take ourselves to have convincingly eliminated or discredited any and all of its proposed rivals' (ibid.). This circumscribes Stanford's scepticism quite a bit: many scientific results do *not* involve a process of eliminative inference: 'the reasons we can give for believing that dinosaurs roamed the earth long ago and that tiny creatures invisible to the naked eye fill the world around us seem non-eliminative in character' (p. 33), and 'Eliminative inference is certainly not the whole of what science does' (p. 32). But Stanford wishes to restrict his scepticism even further than this, since sometimes eliminative inferences are in fact reliable. Indeed, 'In many of the epistemic circumstances we encounter this kind of inference is a perfectly reliable tool'

[5] This is the epigraph of Chapter 1, and is a quote from *Notebooks of a Biologist*, by Jean Rostand.

[6] As is very common in philosophy, and elsewhere, Stanford has succumbed to a (rather irresistible) temptation to make dramatic on-the-surface claims, whilst making much more cautious and circumscribed claims when one digs into the details. I can't claim I have never done this myself.

(p. 29). So the question now is: *precisely which kinds* of eliminative inference should not be trusted?

This is tricky. Stanford writes, 'Often enough . . . we are rightly confident in our ability to have exhausted the space of likely or plausible explanations' (p. 31). And when we are *rightly confident* about this, eliminative inferences are trustworthy and we are justified in believing the explanation that remains once the others have been sufficiently discredited. We should *only* be sceptical when we are 'in a position to reasonably doubt [our] ability to exhaust the space of plausible alternatives' (p. 40). And when, exactly, will this be? If developments in theoretical physics since the turn of the 20th century have taught us anything, it is that we *should* doubt our ability in that context.[7] After all, quantum mechanics makes very little sense to us human beings, containing as it does bizarre conceptual paradoxes that put a wrecking ball through our intuitions. We can't possibly expect to have exhausted all possible alternatives in a context where things that 'don't make sense', or which contain claims that many of us consider plain outrageous, can be serious contenders (e.g. the 'splitting universe' interpretation of quantum mechanics, where 'everything that can happen does happen'[8]).

Naturally, it makes sense for *anybody*, including 'scientific realists', to be extremely cautious about making knowledge claims when it comes to (certain corners of) theoretical physics. Nobody could reasonably say, for example, that we know that the splitting universe interpretation of quantum mechanics is true. There simply isn't sufficient evidence for such a claim, and any realist will be completely happy with that. We may ask, then, for a case where Stanford and the realist really do come apart. Where will the realist say 'the evidence is in', whilst Stanford preaches caution on the grounds that we can't reasonably take ourselves to have exhausted the space of plausible explanations?

An interesting test-case is the asteroid impact theory of the extinction of the dinosaurs (to be analysed in detail in Chapter 7). This was apparently a case of eliminative inference: back in the 1980s, serious 'how-possibly' explanations for the extinction of the dinosaurs included:

1. Climate change
2. Supervolcanoes
3. Asteroid impact

[7] Much more on this in Chapter 6.

[8] Cox B and Forshaw J (2012), *The Quantum Universe: Everything That Can Happen Does Happen*, Penguin.

By 2010, the asteroid impact hypothesis was deemed the frontrunner by the majority, such that an international panel of 41 scientists came together to announce a consensus of opinion (Schulte et al. 2010). The opinion of the panel was *not* merely that the asteroid impact hypothesis was the best of the explanations available, but, rather, that it was reasonable to say that the evidence for its truth as a how-actually explanation was exceedingly strong. It was apparently now reasonable to say that this is something we *know*: an asteroid impact was responsible for the extinction of the non-avian dinosaurs (and much else besides), around 66 million years ago.

What would Stanford say about this? It seems to be a clear case of eliminative inference, so should we consider the possibility that there is some unconceived alternative, and the asteroid impact explanation is merely the best of those we have *currently* conceived? Stanford would apparently have us ask: Can we reasonably take ourselves to have exhausted the space of plausible explanations? Well, we certainly know a lot, these days, about how mass extinction events can occur—there are only so many different ways. And we are talking here about things understandable by human beings—this is *not* like certain corners of non-classical theoretical physics, where we have to question whether we even have the concepts to understand what lies behind the phenomena. So perhaps Stanford would say 'yes', and agree that we know that the asteroid impact theory is (at least approximately) true. And thinking about cases such as this one, one might start to wonder just *where* Stanford and the cautious self-described 'scientific realist' would disagree. They might in fact agree on all of the 30 examples of 'future-proof science' listed in Chapter 1.

What would bring Stanford to believe in the asteroid impact theory? Reading Chapter 2 of *Exceeding Our Grasp*, one might think that Stanford would sign up to the theory when it is clear that we have exhausted the space of reasonable alternative explanations, and we have eliminated all but one. But this can't be right. Scientists in 2010 would *not* have announced a consensus on this theory merely on eliminative grounds. Rather, they made their announcement on the basis of *the quantity and quality of the evidence*, including (just to scratch the surface) the fossil record, the Chicxulub crater (including shocked quartz, sinkholes, etc.), radioactive dating, and sedimentology (e.g. iridium in the K-Pg boundary). Godfrey-Smith (2008, p. 142, fn. 1) writes that Stanford (2006) is 'mostly discussing cases' which have a 'mixed' character, involving both (i) positive support for a theory, and also (ii) comparison with other proposed theories. Both (i) and (ii) were required for the (near-) consensus on the asteroid impact theory. But Stanford provides little detail on (i).

Stanford says much about 'historical evidence' in his book: this is evidence from the history of science that there have sometimes been unconceived alternatives to our best theories. But he says very little about first-order scientific evidence for scientific theories. In particular, he doesn't tell us much about the circumstances in which we might consider the evidence to be very strong, such that it is rational to believe a theory is (at least approximately) true. This is understandable up to a point: Stanford's book is about scepticism, and his case studies (Chapters 3, 4, 5) concern scientific theories which were rejected. But on the other hand it is only possible to understand a sceptical stance on science if one has a good sense of which contemporary scientific ideas the sceptic is in fact happy to believe. And that means talking about evidence, and in particular how to judge when the weight of evidence has crossed some threshold such that it is reasonable to believe the theory, or even such that it is *un*reasonable *not* to believe the theory. This is crucial for a proper assessment of the asteroid impact theory (as will be made clear in Chapter 7).

When I talk about 'weight of evidence', I don't mean to suggest that evidence can be measured, or put onto a linear scale from 0 to 1, where 0 is no evidence, 0.5 is 'medium evidence', and 1 is 'perfect evidence'. No doubt, it is much(!) more complicated than that. I fully agree with Stanford when he writes that there is a 'heterogeneity of forms of evidence, inference, and argument in scientific contexts' (2011, p. 898). Stanford's example, motivating this statement, concerns experimental taphonomy as evidence for the organic theory of fossil origins. In short, these days scientists can recreate fossilisation in the lab, closely mirroring, in a short period of time, how fossilisation over many millions of years would happen, and did happen. Since we know how fossils form, to the extent that we can now form them ourselves,[9] it would nowadays be considered inexplicable if we did *not* find fossils in the way we do (cf. Stanford 2011, p. 895f.).

Stanford is convinced that *this* evidence is conclusive, such that we should believe the organic theory of fossil origins; it supplements the 'abductive evidence' (the organic origin theory is an excellent explanation of the phenomena) we already had (p. 896). Note what Stanford does *not* do in this paper: he does *not* argue that scientists had several theories, were convinced

[9] Stanford cites, for example, Briggs D and Kear A (1993), 'Fossilization of soft tissue in the laboratory', *Science* 259, p. 1439.

that they had exhausted the space of all reasonable candidate theories, and eventually managed to eliminate all of them except the organic theory. It is true that scientists once had several theories, so in a sense an eliminative inference is in play. But Stanford is clear: we should believe the theory because of the overwhelming, first-order scientific evidence for it (with experimental taphonomy playing a major role).

Of course, in many scientific contexts there is nothing at all analogous to experimental taphonomy, so this powerful form of evidence—which Stanford describes as 'projective' in opposition to 'abductive'—is absent. This leaves open for discussion the circumstances in which Stanford would believe a scientific theory on the basis of overwhelming first-order evidence but where such projective evidence is absent. I am left wondering how many of the 30 examples listed in Chapter 1 he would agree to being (probably) future-proof, but there is little in the *details* of his writings to indicate that he would differ from the self-described 'scientific realist' very much (though, as we've seen, there is much on the *surface* of his writings suggesting great disagreement). This is very surprising, and worrying for the debate. Junior scholars in particular will assume that the realism/antirealism debate is a substantial and serious debate between scholars who are very significantly opposed on the question whether we should think that our scientific theories are uncovering the truth about the nature of reality. So let me provide a little further evidence.

Stanford (2015a) makes reference to contemporary scientific theories, and advocates scepticism. Again, on the surface we have a radical form of scientific scepticism (as with Wray 2018), where even all/most of the 30 examples given in Chapter 1 should be considered provisional and apt to be 'rendered obsolete'. But look again.

Stanford in this paper emphasises the important point that sometimes we don't have the continuity of theory that our terminology might suggest. For example, one might insist that we know 'atoms exist', and this *statement* is future-proof, and yet—says Stanford—we may in the future (learning from history) come to change our minds very significantly about what an 'atom' really is, such that the word 'atom' means something very significantly different 1000 years from now. So too, says Stanford, for 'gene':

> [I]t may well be that we continue to hold "genes exist" to be true indefinitely while our beliefs about genes change in just the sorts of profound and unpredictable ways that the historicist critic of scientific realism supposes they will. (2015a, p. 410)

Similarly, one may look back several hundred years, and find scientists of the day saying the words 'atoms exist'. But that—saying the same *words*—is definitely not 'future-proof science' in the sense this book intends.

How does this bear on the 30 examples given in Chapter 1? Very little, I suggest. The cautious, contemporary realist may well agree that scepticism is reasonable when it comes to many of the 'deeper' questions about the nature of atoms, or about the nature of genes. For any particular theory concerning what genes 'really are', the realist may agree that the evidence is not yet sufficient for a confident knowledge claim. Contrast this with the 30 examples. It is most unclear that the arguments in Stanford (2015a) could be applied to any of those examples. Stanford's paper is mostly concerned with questions of reference and meaning (applied to 'atom' and 'gene'), but 29 of the 30 examples in Chapter 1 simply do not face the same problems of reference Stanford so carefully discusses (I exclude example 19). Or at least, not to anything like the same degree. We may well make *some* changes to what we mean by 'tectonic plate' over the next 1000 years. But it isn't feasible that this will so dramatically affect the meaning of contemporary claims about past and ongoing plate tectonics that such claims are not at least approximately true, in a straight-forward sense.

Stanford (2015a, p. 410) does mention 'massive changes' in how we think about chromosomes, planets, and fossils. But he is merely making a point about how we have sometimes, in the history of science, retained a word despite the meaning of that word changing dramatically. At this point in his paper he is not suggesting that we can expect 'massive changes' in the meanings of these words in the future, even if he *does* spend most of the paper arguing that we can reasonably expect such changes for 'atom' and 'gene'. When it comes to the word 'fossil', for example, Stanford thinks we now know what fossils are, how they formed, and where they came from, including many details such as why sea-life fossils are found at the top of mountains, and why many fossils are not exactly like any living creature, as discussed in Stanford (2011). Stanford (2015a) also very briefly mentions 'tectonic plates' (see p. 415f.), but only in his final concluding paragraph, and not in his core arguments.[10]

[10] For me, this is all too close to establishing scientific scepticism using certain cherry-picked cases that make one's arguments as strong as possible ('atoms', 'genes'), then applying the sceptical conclusions broadly, to cases which are significantly different ('tectonic plates'), and which could *not* have been used to support the core arguments. This dodgy dialectical move is widely instantiated, unfortunately, including in the hugely influential *Structure of Scientific Revolutions* (Kuhn 1962).

Stanford's core argument in this paper concerns 'atoms' and 'genes', cases where—at least when it comes to many of the *details*—the contemporary realist herself might harbour concerns vis-à-vis future-proof science. However, Stanford's conclusions sometimes seem to be directed at *all of our best contemporary scientific theories*. For example, Stanford writes (and seems clearly to be asserting) that,

> [O]ur own historical successors will someday view even the leading scientific theories of our own day in very much the same way that we regard those of our historical predecessors. (2015a, p. 412)

Without significant caveats this conclusion is surely an exaggeration, however, since I am reasonably sure that Stanford himself believes in many of our leading scientific theories on the basis of the overwhelming evidence for those theories.

I may have revealed a certain amount of frustration when I read Stanford, but I don't wish to overstate it. I do think that Stanford has done more than most to try to achieve real progress in this debate, separating a realism debate that has reached a stalemate or 'ennui'[11] from (to use Stanford's own words) 'a realism debate that makes a difference' (Stanford 2015b). The most important thing Stanford has done in recent years, in my view, is emphasise just how much 'antirealists' and 'realists' have come closer together, such that one can talk in terms of a 'middle path' in the realism debate (Stanford 2021). Such a 'middle path' concerns a certain amount of agreement between 'antirealists' and 'realists'. On the one hand, the scepticism of antirealists is now typically circumscribed quite significantly, sometimes such that it only concerns cases the realist would *also* be sceptical about. In this way the sceptic acknowledges that there are many scientific results that can be considered items of human *knowledge*, such as when Stanford (2011) argues that the evidence for the organic origin theory of fossils is overwhelming. On the other hand, the optimism of realists is now typically *also* circumscribed, such that many realists expect significant theoretical change, even 'revolutions', in at least some scientific fields in the future. For example, the realist may agree that sometimes there is a *problem of unconceived alternatives*, but not always, and that this problem is most serious for non-classical theoretical physics, or perhaps 'fundamental' physics (see Chapter 6). A realist such as Hoefer (2020) may conclude that no realist should *be* a realist about theories coming

[11] Fine (1986); Stein (1989); Magnus and Callender (2004).

under the heading of 'fundamental physics'. In addition, as Stanford (2020) rightly notes, many realists now only make knowledge claims concerning certain *working parts* of contemporary scientific theories (not the whole theory). This is consistent with major theoretical change in the future (cf. Vickers 2017).

Similarly, Chakravartty (2021) describes a 'realist tightrope', whereby the realist in recent years has been trying to delineate a position that commits her to neither *too much*, nor *too little*, but somehow gets the balance just right. That is, the realist is more cautious than she used to be in the 1970s and 1980s, no doubt: she may renounce realism when it comes to fundamental physics, and, in a given case, she may renounce realism about many theoretical elements which are either 'idle wheels', or somehow can be shown to play a rather minor role (e.g. a merely heuristic role) in delivering the theory's successes compared with other 'working' elements which can be shown to be doing the 'heavy lifting' (McMullin 1984, p. 30; Kitcher 1993, p. 143, fn. 22). But the realist wants to get the balance just right: she doesn't want to commit to too little, such that we can only make knowledge claims in rare cases (e.g. the case of the organic origin theory of fossils discussed above). It must be added, however, that Chakravartty's tightrope is not just a *realist* tightrope: it's an *antirealist* tightrope too. Stanford, for example, has worked hard in various papers over the past 20 years to make sure he is not sceptical about *too much* (e.g. Stanford 2011), nor *too little*. And every 'antirealist' or 'instrumentalist' or other kind of 'sceptic' faces the same dilemma. Consider Van Fraassen, considering a case which clearly concerns *observables* (as he defines that term). Will he believe the scientific claims concerning the existence, properties, and behaviour of those observables? Well, he will have to judge the weight of evidence somehow.

The upshot of all this is that we really are all in this together, trying to establish quite precisely the circumstances in which it is reasonable to be sceptical, and the circumstances in which it is *not* reasonable. Stanford and I seem at first like great adversaries in the *science and truth* debate: the titles of our books, *Exceeding Our Grasp* and *Identifying Future-Proof Science*, pull in opposite directions, and he has self-described as an 'instrumentalist' whereas I have (in the past) self-described as a 'realist'. But behind the 'isms', the headlines, and the soundbites, there is very considerable agreement.[12]

[12] It's worth noting that I *am* in serious disagreement with the early Van Fraassen (1980), who put so much weight on *naked eye* observation. On that view, several of my 30 examples would be denied. I consider this a very fringe view nowadays, and even Van Fraassen has left it behind. Dellsen (2017,

3. The Historical Challenge: Are We Epistemically Privileged?

The pessimistic induction (made famous by Laudan 1981) uses several successful-but-wrong theories in the history of science in an attempt to sever the link between success and truth, with the conclusion that we should not believe in contemporary theories on the basis of their success. But consider contemporary theories concerned with *unobserved observables* (such as the core of the Earth) that the vast majority of self-described 'antirealists' agree we have great warrant to believe in. Where does the warrant come from? It comes from 'success', broadly construed, especially as perceived by those scientific communities that have formed a strong consensus around those theories. In other words, when it comes to scientific claims concerning observables—e.g. 'The Sun is a star'—even *antirealists* want to make a success-to-truth inference. It might be reasoned: it would be exceedingly unlikely (informally speaking, a 'miracle') for our scientific observations to accord so perfectly with the assumption that the Sun is a star if the Sun is *not* a star. Thus, it is *highly* likely that the Sun is a star. If the antirealist will accept this, then she finds herself needing to offer her *own* response to a slightly modified 'pessimistic induction' where all of the examples of successful-but-rejected theories involve claims concerning 'observables' (broadly construed). In which case, the 'realist' and 'antirealist' might as well just work together; they both equally need to articulate the reasons why we can be so confident in the (approximate) truth of many of our current scientific claims concerning 'observables' given the eventual demise of past—yet successful—theories concerning 'observables'.[13]

Many realists faced with the historical threat wish to point to very significant differences between past and present science.[14] A great many scientific publications of times past quite clearly fail to meet the standard of rigour and general professionalism that is nowadays required for publication in even a low-ranked science journal. One might simply say: 'We do science better now!' And indeed, it has been measured that—given the quite dramatic exponential growth of science (Figure 2.1)—95 per cent of all science ever done has been

p. 108)—drawing on some of Van Fraassen's more recent work including Van Fraassen (2001)—notes that '[F]or van Fraassen, what is observable is itself a matter of empirical investigation, and thus there is no simple rule for telling what counts as observable or unobservable.'

[13] A similar point could be made about Stanford's 'new' induction, based on unconceived alternatives. If this argument really threatens realists, then it also threatens all antirealists, excepting the vanishingly small minority of radical antirealists who would even doubt that the Sun is a star (say).

[14] See e.g. Alai (2017); see also Fahrbach (2021), fn. 2.

done in the past 100 years (see also De Solla Price 1963). One might well expect that in the first 5 per cent of any project mistakes are likely; in the first 5 per cent one is 'just getting warmed up'. Fahrbach turns this into an explicit defence against the pessimistic induction. If we take a simple, linear perspective on the history of science, a pattern of 'revolutions' in our theory of light seems to suggest further revolutions in the future (Figure 2.2). But, says Fahrbach (2011, p. 150), '[T]he bare amount of time has nothing to do with the degree of success of theories, so the linear weighting is implausible.' Given Figure 2.1, what we really need is an exponential weighting (Figure 2.3). On this model, a similar amount of

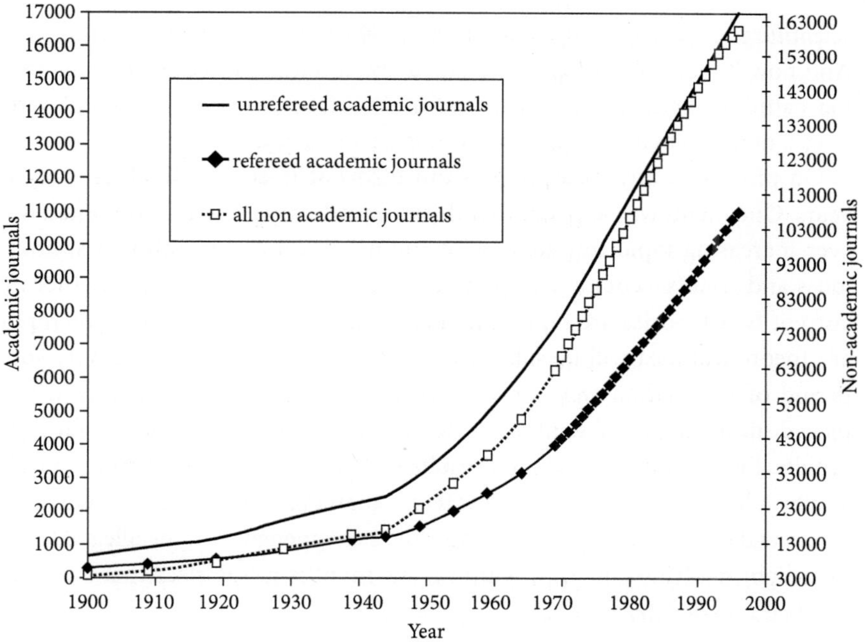

Figure 2.1 The exponential growth of science during the 20th century

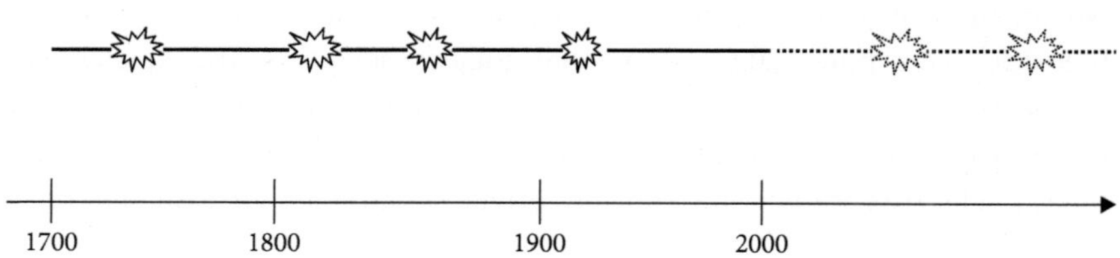

Figure 2.2 Revolutions in our thinking about the nature of light, with a linear x-axis, and including an indication of how one might expect the pattern to continue

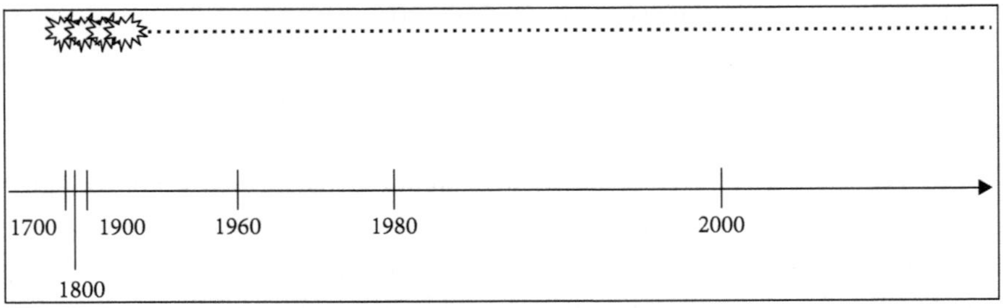

Figure 2.3 Revolutions in our thinking about the nature of light, now with an exponential x-axis corresponding to the exponential growth of science

scientific effort occupied the years 1700–1980 as occupied the years 1980–2000. And now it looks like, whereas we had a few revolutions in our thinking about the nature of light when science was in its infancy, we have since enjoyed a longstanding stability, perhaps suggesting permanency.

On just about any measure you can throw at it, science doubles every 20 years. One might well suppose that this extraordinary growth—combined with ever-increasing rapid international communication and scrutiny of scientific ideas and results—couldn't help but be relevant for the question whether our current best theories are (approximately) true. For example, any contemporary theoretical idea will be subject to far more scrutiny—in a relatively short period of time—than were theoretical ideas of the past. Any fallacy will be picked up quicker, and communicated to every corner of the world far more rapidly. Significant experimental results will be double-checked more quickly, and by a larger number of scientific teams than ever before. Any experimental errors will likely be identified very quickly, as happened with the alleged faster-than-light neutrinos of 2011; within a few months it was clear that a mistake had been made (Reich 2012).

This argument from exponential growth has been quickly dismissed by some antirealists. Consider Fahrbach (2011, p. 149) insisting that, 'more scientific work results in the discovery of more phenomena and observations, which, in turn, can be used for more varied and better empirical tests of theories.' This potentially means that today's scientists are 'epistemically privileged' compared with previous generations of scientists. But Wray (2018, p. 65)—drawing on Hesse (1976)—advocates 'the thesis of no privilege': today's scientists are *not* epistemically privileged after all. He explains (p. 94):

> The realist should realize that every generation of scientists (and philosophers of science) could run an argument similar in structure to the argument developed by Fahrbach.

In other words, the argument from exponential growth can't be a good argument, because past scientists could *also* have appealed to exponential growth, saying things such as 'Science is now bigger and better than ever before!' and identifying a significant period of theoretical stability, as in Figure 2.3. And yet, as we now know, many of their theories came to be rejected. To give one concrete example, physicists in the year 1900 could certainly have said 'Physics is bigger and better than ever before' and 'Science has grown exponentially.' And they could have pointed to long-ago revolutions followed by a very significant period of stability/knowledge accumulation. Frost-Arnold (2019, p. 908) makes the same point vis-à-vis scientists in 1850 looking back on scientists in 1720.

So far we see the usual realism debate dichotomy: the realist Fahrbach defends against the pessimistic induction, and the antirealist Wray attacks that attempted defence. However, the realist, instead of offering her own defence, might instead ask the antirealist why *she* believes in certain scientific results that concern unobserved observables. For example, all of the non-radical 'antirealists' would agree that we know that the Sun is a star. And yet, we can't *see* that the Sun is a star. So why do they believe this? At this point it seems inevitable that the antirealist will appeal to scientific evidence. But it is here that she meets with her own arguments. If Wray really wants to insist that we are *not* epistemically privileged compared with previous generations of scientists, then our own appeal to evidence is no more reliable than *their* appeals to evidence. Kelvin thought he had very good evidence—based on thermodynamics—that the Earth could not be more than 100 million years old, but he was wrong, and in a big way. Any belief in a contemporary scientific claim, based on scientific evidence, will necessarily depend upon some kind of distinction between the present evidential situation and past evidential situations. It's hard to see how Wray and Hesse's 'thesis of no privilege' can be squared with both (i) the history and science, and also (ii) the thought that scientists *really know* that the Sun is a star.[15]

Wray is right that, if we want to find a distinction between past and present science with great significance for epistemology, we can't *merely* point to how much better science is now. Past scientists could have said the same thing, pointing to an even earlier time. In particular, we can't *simply* say that the evidence we have, for our theories, is much better than the evidence they had, since once again they could have said the same thing about an even earlier

[15] If the reader is worried about the dialectical weight I am placing on 'The Sun is a star', she might usefully consult Section 2.4 of Chapter 9, where the statement is unpacked. Alternatively, there are many other examples that could be used.

generation of scientists: 'Our evidence is *much* better than the evidence *they* had.'

Fahrbach makes a couple of mistakes. First, we don't exactly want a distinction between past and present science, as if past scientists never had sufficient evidence to make a knowledge claim, and we (sometimes) do. In fact, scientists of the past sometimes *did* have sufficient evidence (e.g. to assert that the Earth spins on its axis), and contemporary scientists often *don't* have sufficient evidence (e.g. to assert that Tiktaalik is a true missing link—see Chapter 4; or, perhaps, to assert that an asteroid killed the dinosaurs—see Chapter 7).

Second, we don't want cases of evidence sufficient for *some* leading scientists (such as Kelvin) to assert a knowledge claim. Nor are we looking for cases where there is a scientific consensus as to which theory is *the current front-runner*. Instead what's required is for there to be a solid scientific consensus that the evidence is sufficient to make a knowledge claim. And we should *still* not be impressed if the relevant scientific community is very small, and/or lacks any diversity of perspectives. In identifying four revolutions in our theory of light (Figures 2.2 and 2.3), Fahrbach makes relevant cases that should *not* be relevant. For example, if we consider the Newtonian, corpuscular theory of light, there was never anything like an international scientific consensus, even during its heyday in the 18th century.[16] Not in anything like the way we find an international scientific consensus concerning the 30 examples listed in Chapter 1.

The basic idea is inspired by Oreskes (2019). She asks the question 'Why trust science?' and her answer is rooted in a certain kind of scientific consensus. As she notes, 'Scientific consensus is hard to come by' (p. 128). This is clearly evidenced by the history of science; for example, it has been well documented that the fight for atomism was incredibly hard won (e.g. Gardner 1979; Chalmers 2009; Ivanova 2013). After all, any sufficiently large and varied scientific community will include within it a wide range of opinions and attitudes (see Chapter 5 for detailed discussion). Reaching a 'solid' (at least 95% agreement) scientific consensus thus requires enough evidence to convince those that are initially inclined to be (very) sceptical. This makes the size and diversity of the scientific community important—a consensus is going

[16] Leading scholars of the day including Leonhard Euler and Benjamin Franklin objected to the theory, and the Newtonians defended it; see e.g. Horsley (1770) and Cantor (1975, p. 115). There was arguably a consensus *within Britain* during the 18th century, but localised consensuses are something to be wary of, as will be further discussed in Chapter 5 in the context of the continental drift controversy.

to be harder to reach the greater the size and diversity of that community. Thus, in her pursuit of trustworthy science, Oreskes (p. 143) looks not only for a scientific consensus of qualified experts, but in addition requires the relevant scientific community to include within it 'a range of perspectives'. A good route to achieving this, she argues, is to have demographic diversity in the community, where we might consider age, gender, race, ethnicity, sexuality, and country of origin (p. 144). It must also be the case that there has 'been ample opportunity for dissenting views to be heard, considered, and weighed' (ibid.).

By requiring this kind of scientific consensus for future-proof science it becomes hard to mount any kind of historical challenge. For example, past scientists sometimes thought they had good evidence for theories that have now been rejected, it is true—but they never formed a 95 per cent international scientific consensus explicitly declaring something to be 'scientific fact' that was later to be rejected (as opposed to adjusted).

All of this is not to say that the exponential growth of science and the 'explosion of evidence' Fahrbach (2011, 2017, 2021) discusses is irrelevant. The exponential growth of science brings with it an increase in the size and diversity of perspectives of the scientific community, and this makes any international scientific consensus all the more significant. The fact that there are *any* international consensuses is remarkable, and would be inexplicable were it really the case that the evidence was not inordinately strong. At some point, it becomes absurd to suggest that future generations will look back on *us* and say, 'They thought they had very good evidence for their beliefs, but now we know better.' At some point this is a possibility only in the same sense that it is possible we are all hooked up to The Matrix. And antirealists who accept that we really *know* that the Sun is a star will want to agree.

4. Weight of Evidence Judgements: Scientists vs Philosophers

I've argued that the antirealist, just as the realist, needs to *weigh up the evidence* when she asks herself which contemporary scientific theories she is going to believe in. Decades of scholarship shows just how extraordinarily difficult this often is. For one thing, evidence comes in many forms, as already noted (and as we'll see in the case studies). The evidence coming from experimental taphonomy for the organic origin of fossils is quite unique (Stanford 2011) and significantly different to the types of evidence at play in most other scientific contexts. There is a real sense in which every case is different, and generalisations are dangerous.

If generalisations are dangerous, we need to 'go local' (Asay 2019; Saatsi 2017; Henderson 2018). And if we take 'going local' seriously we may worry that all we can do when it comes to gauging the weight of evidence for (different parts of) a theory is ask the scientists. It is reasonable to suppose that the vast majority of evidence for/against a given scientific theory is going to be first-order scientific evidence (Fitzpatrick 2013). And scientists know much more about this, in a given case, than do philosophers of science. So when we ask 'What is the weight of evidence?', we simply need to go and ask a scientist. Or, better, we should canvass the opinions of a large number of scientists, disregard the outliers (those who are biased, have an agenda, etc.), and see what kind of consensus (if any) we can identify in what remains. And if the vast majority of scientists think the evidence is extremely strong, then we might reasonably infer that the evidence *is* extremely strong.

This isn't too far from the best we can do—so this book will argue. Various scholars have noted scientists making false proclamations in the history of science. We already saw in Chapter 1 that Lord Kelvin, at the turn of the 20th century, reportedly stated that 'There is nothing new to be discovered in physics now. All that remains is more and more precise measurement.' And Albert A. Michelson (famed for the Michelson–Morley experiment of 1887) wrote in 1903:

> The more important fundamental laws and facts of physical science have all been discovered, and these are so firmly established that the possibility of their ever being supplanted in consequence of new discoveries is exceedingly remote. (Michelson 1903, p. 23f.)

One might also consider James Clerk Maxwell, who is said to have described the (now rejected) aether as '... better confirmed than any other theoretical entity in natural philosophy' (Laudan 1984, p. 114). However, scholars have *not* found examples of entire *communities* of scientists making such false proclamations. At least, not in the last 95 per cent of all science ever done. And the above quotes are embarrassing only because of revolutions in 20th-century theoretical physics, which is to some extent a special case (see Chapter 6 for discussion).

Thus I do now think I made a mistake with my *The Conversation* article of 2018. This article was entitled 'The Misleading Evidence that Fooled Scientists for Decades'. But I now think only a small number of scientists were ever 'fooled for decades', especially when we look for examples outside theoretical physics. My first example in that article concerns Meckel's successful prediction of gill slits in the human embryo, on the basis of a radically false theory of biological development. But this did not mislead the scientific community of the day, it turns out, even if it might initially seem as if it should have (see Chapter 3).

In fact, this book will argue that the history of science is consistent with the following very bold claim: the international scientific community is now (and has been for at least 100 years) *infallible* at judging when the weight of evidence for a particular scientific idea is sufficient to make a knowledge claim. The reader may immediately start thinking of possible counterexamples, where the international scientific community *was*, in the last 100 years, 'sure' about something, and then they had to change their minds. A suggested counter-example concerns attitudes *against* Wegener's continental drift theory. Indeed, one may perhaps compare two conferences: (i) the 1939 Frankfurt conference on the history of the Atlantic, and (ii) the 41st Lunar and Planetary Science Conference held in 2010. At the former, there was quite a significant consensus of those gathered *against* continental drift, and in favour of continental fixity. In hindsight the 'consensus' view was thoroughly mistaken—continental drift is now a fact. At the latter conference, in 2010, an apparent international consensus was agreed that an asteroid impact was the major cause of the K-Pg extinction event (the extinction of the non-avian dinosaurs). But how could we possibly consider this opinion on the extinction of the dinosaurs to be future-proof, given that the opinion on continental fixity wasn't?

The continental drift case will be handled in detail in Chapter 5. It is particularly important, and deserving of a full chapter, given the way it apparently undermines claims of future-proof science which appeal to 'scientific consensus'. Certainly this case shows that the advice 'just ask a scientist' can go wrong. Whilst at Cambridge University in 1946/7, studying geology and zoology, Sir David Attenborough recalls that he *did* 'ask a scientist' about continental drift. He recalls:

> At university I once asked one of my lecturers why he was not talking to us about continental drift and I was told, sneeringly, that if I could prove there was a force that could move continents, then he might think about it. The idea was moonshine, I was informed.[17]

[17] See the 2012 *The Observer* article 'David Attenborough: force of nature', available at https://www.theguardian.com/tv-and-radio/2012/oct/26/richard-attenborough-climate-global-arctic-environment (last accessed February 2022).

With stories such as this one in hand, we may worry about trusting scientists' intuitions. And indeed, one might already have expected this; it is well known that our intuitions—despite being impressive in some respects—let us down in all sorts of ways. Why should they be reliable when judging something as complex and multifaceted as the overall weight of evidence for a scientific claim, hypothesis, or theory?

A serious-minded philosopher might well insist that we need to leave intuitions behind, and that evidential judgements need themselves to be 'made scientific' by *quantifying* them. And indeed, through the 20th century a lot of energy was poured into quantitative theories of evidence, most commonly referred to as *theories of confirmation*. The front-runner was Bayesian confirmation theory, and this has remained popular for many decades now. A tantalising suggestion presents itself: perhaps philosophers of science can contribute something of vital important to science, by devising a quantitative theory of evidence which can be used to measure the weight of evidence for a scientific claim, which is much more rigorous and reliable than the mere intuitions of scientists, and which is immune to various well-known weaknesses in human reasoning (and scientists are human too!) including confirmation bias, the base rate fallacy, groupthink, the look-elsewhere effect, and others too numerous to mention.

However, if Bayesian confirmation theory was ever thought of as a way to judge the *overall* weight of evidence for specific *scientific* claims/hypotheses/theories, it has failed. Indeed, modern comprehensive treatments such as *Bayesian Philosophy of Science* (Sprenger and Hartmann 2019) don't even attempt a serious Bayesian analysis of any particular scientific claim. The book is very thin on science, in fact, with more energy put into simple, toy examples such as pulling balls out of an urn.[18] This is typical of the Bayesian literature: whilst the discussion is constantly *directed at* science, one hardly ever finds a serious Bayesian scientific case study, and often even 'toy' case studies are absent. The closest Sprenger and Hartmann (2019) come is a mini-case-study of a particular experiment in parapsychology (see pp. 300–6). However, the point of this case study is *not* to use Bayesian techniques to judge the evidence, and determine what we should believe about the relevant claims. The point is

[18] Most work in Bayesian philosophy of science is directed *at* science, but thin on science. A good recent example is Landes (2020). The technical discussion found here is apparently directed at understanding an important aspect of the nature of scientific evidence, but the gap between the highly abstract discussion in this paper and actual scientific practice is huge. Rather than contributing something to our understanding of how science works, pieces such as this demonstrate how Bayesians are in danger of getting tangled up in their own self-made problem spaces.

much more modest: to show that 'the subjective Bayesian framework enhances transparency in statistical reasoning'. Very modest 'mini' case studies such as this can be found in the Bayesian literature (cf. Dorling 1992), but they can't help with the specific goal of this book—simply, to determine what we should believe. It might yet be suggested that Bayesian ideas cannot *pre*scribe, but can merely *de*scribe—perhaps just roughly, crudely—what is going on when scientists judge evidence. But even this more modest aim seems hopelessly out of reach. Kitcher (2002) makes the central point when he writes, 'People don't actually accept the main claims of chemistry (or any other science) because they have approximated some Bayesian procedure' (p. 396f.).[19]

This is prima facie surprising. Why *wouldn't* a carefully crafted Bayesian epistemology—developed by philosophers over several decades—do a better job at weighing evidence than the intuitions of scientists, human beings who are susceptible to all those fallacies of reasoning? One reason is that the fallacies themselves hardly touch the overall scientific community at all. If an individual scientist or team of scientists commits the base rate fallacy (say), it only takes one scientist in the world, amongst all the millions of scientists, to call them out on it, and in very little time the whole world will know they committed the fallacy. Similarly with confirmation bias, and a host of other fallacies one could name. But anyway, most fallacies don't even appear in published scientific papers these days, since every scientist worth her salt is aware of relevant fallacies, or else they are caught at the peer-review stage. One might point to the 'replication crisis' as a context where fallacies were committed in published articles, and Bird (2018) even argues (convincingly) that the base rate fallacy played a role here. However, the replication crisis is hardly a crisis. Although a handful of scientists might have been convinced by some of the published results which it turns out are not replicable, the evidence for these results was always extremely thin, often consisting of a single isolated study.[20] Hardly grounds for an international consensus. In this book my interest is in cases where the evidence is far stronger; after all, it is only in cases of very strong evidence that we will find future-proof science.

[19] Various scholars are keen to stress the limits of Bayesian confirmation theory. On this issue I am in agreement with Cartwright et al. (forthcoming): good science involves a complex tangle of many different elements, a fact that doesn't sit well with a Bayesian approach to confirmation of hypotheses, where one simply plugs in evidence and calculates the posterior probability. To use Cartwright et al.'s favourite analogy, one doesn't come to understand how the Jacana bird's eggs are supported by considering each leaf, stem, and twig individually; it's the tangle of all these things together that matters.

[20] This point is made by Oreskes (2019, Chapter 7): 'Most of the studies that Professor Krosnick offers as evidence of trouble in science are single studies that were later shown to be faulty [. . .]. We should be skeptical of any single paper in science.'

So Bayesian confirmation doesn't improve on scientists' intuitions when it comes to avoiding fallacies—scientists are already very good at that. But what about the fact that Bayesian confirmation is quantitative, and avoids intuitions? Aren't intuitions demonstrably unreliable in a wide range of contexts? But again, we are talking about the overall scientific community, and not individual scientists. It would certainly be wrong to claim that we can trust the intuitions of any given *individual* scientist. There are plenty of examples in the history of science of the intuitions of scientists being seriously opposed. Famously, intuitions concerning the infinitesimals of the early calculus varied dramatically.[21] So too, intuitions during the quantum revolution varied dramatically: some scientists (including Einstein) considered Bohr's successes of 1913 excellent evidence for his model of the atom, whilst others remained unconvinced.[22] We'll come across other examples in the case study chapters. But somehow, during the push and pull of international scientific communication and scrutiny, after all the checks and balances have been performed, we reach some happy medium. And it is much harder to claim that this averaged-out judgement of the evidence, coming from the scientific community as a whole, is unreliable. It may seem odd to speak of a community making a judgement, but, as Godfrey-Smith (2008, p. 142) writes—citing Kuhn (1970), Hull (1988), Kitcher (1993), and Strevens (2003)—'We have become used to the idea that a community or population can embody epistemic properties that no individual has.'[23]

We'll see in the case studies just how reliable this averaged-out intuition concerning the weight of evidence in fact is (or at least has been in past cases). Suppose for now it is very reliable, in most scientific contexts. At this point in our discussion we've lost our reasons for thinking Bayesian confirmation theory is an improvement on scientists' intuitions, but this is just the beginning. There are plenty of reasons for thinking that Bayesian confirmation is much *worse* than the averaged-out intuition of the scientific community. Supposing that Bayesians have good answers to the problem of old evidence and the problem of the priors, scientific evidence is just far too complex and multifaceted to put through the Bayesian machinery, even if we could agree on how to do that. Consider, for example, a Bayesian attempting to fit the

[21] I've discussed the early calculus in detail elsewhere; see Vickers (2013b), Chapter 6.

[22] See Kragh (1985) for a wide range of reactions to Bohr's 1913 model of the atom and the subsequent developments. On hearing of Bohr's success with the ionised helium spectral lines, Einstein is reported as stating, 'The theory of Bohr must then be right' (cited in Pais 1991, p. 154).

[23] There is a wealth of literature on this topic; see e.g. Niiniluoto (2020) and the references therein.

taphonomic evidence for the organic origin of fossils into a Bayesian mould.[24] As Stanford (2011, pp. 897–8) argues,

> [Bayesianism] treats the discovery that actual organic remains deposited in favorable circumstances mineralize as evidence on which a comparatively higher likelihood is conferred by the hypothesis of organic fossil origins than by its negation or by (serious) rival hypotheses. But contemporary thinkers are fully able to appreciate the force of this taphonomic evidence despite the fact that they (including you, dear reader, just a short time ago) cannot articulate even a single serious theoretical alternative to the hypothesis of organic fossil origins and despite the fact that it remains an utter mystery why the negation of the hypothesis of organic fossil origins should confer any particular likelihood whatsoever on processes such as water sorting or the mineralization of organic remains. The austere Bayesian apparatus does promise to allow us to formally integrate the confirmational significance of various diverse forms of evidence, but this remains a promissory note when we have no way to responsibly determine likelihoods of the sort with which taphonomic investigation seems largely unconcerned.

The Bayesian who insists on shoe-horning this case into a Bayesian mould is unreasonably resistant to the fact that scientific evidence comes in many different forms. And a full and proper account of evidence should embrace that, not try to make out that all evidence works in the same way.[25]

This is definitely not to say that Bayesian confirmation is useless. Indeed, I'll use Bayesian ideas to help understand specific instances of evidence in later chapters. Bayesian ideas—including the Bayesian formula for updating credences—are useful tools, to apply when and where appropriate. As a tool, the Bayesian formula can help us to navigate certain fallacies, for example. It can also help us see just how (in)significant an individual piece of evidence is. This isn't about being anti-Bayesian; it's about recognising its limitations. The dream (if anyone ever had it) of analysing evidence in science using a Bayesian

[24] Norton (2021) makes essentially the same point; see e.g. p. 38.

[25] One reason evidence is heterogeneous is that scientific method is heterogeneous. As Oreskes (2019) notes, 'There is now broad agreement among historians, philosophers, sociologists, and anthropologists of science that there is no (singular) scientific method, and that scientific practice consists of communities of people, making decisions for reasons that are both empirical and social, using diverse methods.' Cf. also Hoefer (2020), p. 20: 'The idea that there is a thing called "the scientific method" which is applied in all (or most) sciences and which leads us inevitably to the truth (or closer and closer to the truth) is one that I gave up a long time ago. There are many different methods that can count as scientific, and different ones apply at different times, both within a single scientific domain and when we look across more than one domain.'

formula, to reach a judgement concerning the *overall* weight of evidence for a specific scientific claim, and that can correct the (intuitive) judgement of the scientific community, is dead.

Thus the focus of this book, when it comes to judging evidence, lies more with the intuitions of scientists than with the theories of philosophers. I am thus following in the footsteps of 'naturalists' such as Fine (1986) and Maddy (2001). A well-known statement of Fine's naturalism is, '[O]ne's ontological attitude towards ... everything ... that might be collected in the scientific zoo (whether observable or not), [should] be governed by the very same standards of evidence and inference that are employed by science itself' (Fine 1986, p. 150). In other words, if we ask 'Are electrons real?' we should only consider the first-order scientific evidence. This obviously suggests that we should ask the scientists—those trained to produce and assess the first-order scientific evidence—and not the philosophers of science, who often have little or no formal training in the relevant scientific methods and theories. Whilst I certainly have sympathy for such 'naturalism', there is a curious twist in the tale lurking hereabouts (spoiler alert). Whether the resultant position is still 'naturalism' is a bad question, since that term is so contested and plural (Bryant 2020). To be blunt, this book will argue that one does better to *ask a philosopher* (!) when it comes to the question whether a specific scientific claim is future-proof. This result was a big surprise to me, and thoroughly unexpected when I started writing this book. That is of course one famous definition of philosophy: we start with what seems obvious, make apparently solid inferences, and end up somewhere astonishing.

The long story is important, but the short story is perhaps worth summarising briefly here. It starts by noting that even hardworking scientists only have time to properly consider a small fraction of the overall first-order evidence relevant to any given scientific claim. But we want to base our conclusions on the *entire* evidence base, ideally judged from a variety of different perspectives.[26] The only option is to *in*directly access the overall evidence by assessing the judgement of the *relevant scientific community* (since that community, taken as a whole, *has* directly assessed all of the first-order evidence from a variety of different perspectives). But the philosopher-sociologist will often be better placed than the scientist to *impartially* assess the overall judgement of the relevant scientific community. After all, scientists on

[26] As proposed by Oreskes (2019), building on Longino (2002, pp. 131–2), in turn building on Longino (1990), especially her concept of 'transformative criticism'. For critical discussion of Longino's concept of 'transformative criticism', see e.g. Milliken (2017).

the front line have spent their entire career focused on first-order evidential judgements, and will inevitably have their personal biases and perspectives. Thus their opinion may *not* be representative of the overall community judgement (and we'll see plenty of examples in later chapters). The point is easy to illustrate with an example: consider the case of David Attenborough at Cambridge in the later 1940s, asking about the theory of continental drift. He asked his lecturer, a relevant expert in the field, and his lecturer was keen to impart his personal 'wisdom' of the claim in question. We all know the fate of that 'wisdom'. Had Sir David instead asked a philosopher-sociologist concerned with assessing the overall community opinion, he would have been advised of a divided community (see Chapter 5).

This is just to give the reader a taste of what is to come. Much of what I claim here will be revisited and critically analysed much more carefully in later chapters. The case studies are particularly important as tools to substantiate some of my central claims. Let us now turn to the first of those case studies.

3

Meckel's Successful Prediction of Gill Slits

A Case of Misleading Evidence?

1. Introduction

Can we make an inference from the success of a scientific theory to its probable truth? In his famous paper, Laudan (1981) argued that there are many examples in the history of science of radically false theories being successful. For some, these examples seriously undermined the realist claim that we can make an inference from scientific success to even the *approximate* truth of the scientific theory in question. One of the most important ways in which realists responded was to insist that we need *predictive* success before we are justified in making a realist commitment. Witness Musgrave (1985, p. 211): '[F]ew, arguably none, of the theories cited [by Laudan 1981] had any novel predictive success.'[1] And when Stanford (2006) presented various case studies from the history of biology as challenges to the scientific realist, Votsis (2007) replied in a similar spirit, noting that Stanford focuses on 'theories that only offer explanations'. Saatsi (2009, p. 358) likewise criticised Stanford for focusing on cases which 'by and large do not involve any novel predictive success!'

The prediction of an entirely new phenomenon, which is then tested and vindicated, is certainly noteworthy. There are many examples in the history of science of scholars remarking on the evidential significance of a successful prediction. The thought goes back a long way: William Whewell wrote in 1849 that 'most scientific thinkers... have allowed the coincidence of results predicted by theory with fact afterwards observed, to produce the strongest effects upon their conviction' (Whewell 1849, p. 59; see Bolinska and Martin 2021). Scientists have themselves occasionally waxed lyrical about the power of successful predictions. Consider the following, courtesy of John Playfair, commenting in 1805 on a phenomenon predicted by James Hutton's theory of the Earth:

[1] McMullin (1984, p. 30) noted the importance of novel predictive success a year earlier.

Identifying Future-Proof Science. Peter Vickers, Oxford University Press. © Peter Vickers 2023.
DOI: 10.1093/oso/9780192862730.003.0003

> On us who saw these phenomena for the first time, the impression made will not be easily forgotten. The palpable evidence presented to us, of one of the most extraordinary and important facts in the natural history of the earth, gave a reality and substance to those theoretical speculations, which, however probable, had never till now been directly authenticated by the testimony of the senses. We often said to ourselves, What clearer evidence could we have had of the different formation of the rocks, and of the long interval which separated their formation, had we actually seen them emerging from the bosom of the deep?
>
> (Playfair 1805, p. 72; see Rossetter 2018, p. 8)

Numerous other such quotes could here be provided, from every corner of science (see e.g. Musgrave 1974). Clearly predictions can sometimes have tremendous power to persuade, and Rossetter goes on to note that some of those who had 'previously rejected Hutton's theory entirely, became converted as a direct result of these successes' (2018, p. 8).

However, the story of successful predictions is far from simple. Indeed, within the history of science there are a great many examples of quite *un*impressive 'novel predictive success'. For example, a prediction (though novel) can be very vague, such that when it is successful it is *not* very impressive and should influence our beliefs very little, if at all.[2] In addition, a successful prediction may emerge from within a theory, but where only certain aspects of the theory are really doing the work to generate the prediction. In which case, the success of the prediction bears not at all on the truth of the *whole* theory. The latter point has been widely discussed in the literature: theories can sometimes be divided into 'working parts' and 'idle wheels' (at least relative to particular successes). This allows for a radically false theory—a theory that is not even approximately true—to be successful due to the true elements within it.

Suppose we insist that for a prediction to be evidentially impressive it should be specific (not vague), and thereby *risky* in the sense that it seems highly unlikely the prediction could be successful if the underlying ideas are false (cf. Popper 1962, p. 36; Mayo 1991, p. 528). Suppose too that we take care to confer confirmation only upon those theoretical ideas really doing the work to generate the prediction. Can we then identify future-proof science, by first

[2] Vickers (2013a), pp. 195–8; Vickers (2019), pp. 578–81. In 1991 Deborah Mayo already emphasised that novel predictions do not always carry significant confirmatory weight (Mayo 1991).

identifying scientific theories that enjoy *risky* novel predictive success, and then selecting only the *working parts* of the theory as 'future-proof'?

Even with these caveats, there are examples from the history of science which go firmly against an inference from such empirical success to theoretical truth. If we focus only on novel predictions, several examples on Laudan's list are left behind, but not all (*pace* Musgrave): briefly: (i) caloric theory predicted the speed of sound in air; (ii) phlogiston theory predicted that heating a calx with 'inflammable air' would turn it into a metal; (iii) the aether theory of light predicted the famous Poisson/Arago 'white spot'. New examples, sensitive to the caveats, have materialised in recent decades.[3] And no doubt there are new examples waiting to be introduced (see Lyons and Vickers 2021). A prima facie striking example concerns JF Meckel's 1811 prediction that gill slits would be found in the embryos of 'higher mammals', including humans. His prediction was confirmed via observations that took place between 1825 and 1828.[4] But—with the benefit of hindsight—his theory was radically false. Stanford (2009, p. 384) writes that this case 'illustrate[s] that novel predictive success is no proof of either the approximate truth of a theory that enjoys it or the absence of fundamentally distinct unconceived alternatives to that theory'.

Thus the case of Meckel is surely helpful when we consider the circumstances in which we might label some theoretical idea 'future-proof'. Future-proof science is linked to extremely strong evidence, and novel predictive success has often been described—by philosophers and scientists alike—as the strongest type of evidence there is. Even more so once we introduce the caveats noted above. We may ask whether it is possible for theoretical ideas to result in such success, and such that scientists are persuaded, only for it later to turn out that the evidence was *misleading* (Vickers 2018b; Frost-Arnold 2019). Concrete cases of such misleading evidence are important lessons from history that should influence our willingness to describe any contemporary idea as 'future-proof'.

2. The Gill Slit Prediction: Success from Falsity?

The theory in question for our purposes is the *theory of recapitulation*, and JF Meckel (1781–1833) was a major figure advocating this theory. The theory

[3] See e.g. Lyons (2006); Saatsi and Vickers (2011); Pashby (2012); Díez and Carman (2015); Vickers (2020a).

[4] Modern embryologists tend to refer to 'pharyngeal arches', but they are sometimes still called 'branchial arches', where 'branchial' comes from a Latin word for gills.

requires a basic background assumption, which had been widely assumed for hundreds of years:

> *Scala Natura*: There exists a scale of perfection in nature (the 'scala natura', or *Tierreihe*), from the most basic, most 'imperfect' organisms (e.g. slugs), to the most 'perfect' (human beings).

In addition, it was common sense to the relevant communities that the development of an individual human being passes from something incredibly small, basic, and 'imperfect' to the final 'perfect' end state. This provides us with two series that can be compared: the 'scala natura' series and the developmental series of an individual human being. It is natural to compare them, since they have exactly the same end state (the perfect human being), and they have very similar (possibly identical) starting points. It doesn't take much imagination to suppose that there is a parallel, that the two series are in fact very similar all the way through from start to finish.

It was widely imagined, in fact, that there is an underlying reason for this: perhaps a biological force is responsible for all organic development, but the force is less powerful in the 'lower' organisms, and more powerful in 'higher' organisms such as human beings. The French embryologist Étienne Serres was one major figure who discussed such a biological force explicitly (see e.g. Russell 1916, p. 80). The German naturalist Carl Friedrich Kielmeyer referred to a 'power' common to the two series (see Gould 1977, Chapter 3). Serres wrote in 1830,

> Since the formative force, whatever it is, has less energetic impulse [in lower animals] than in higher animals, the organs run through only a part of the transformations that they undergo in superior creatures. From this it follows that they offer to us, in a permanent manner, the organic configurations that are only transitory in the embryo of man and the higher vertebrates.
>
> (See Gould 1977, p. 49)

On this view, then, 'lower' organisms develop in the same biological direction as human beings, but their development halts at some point, leaving them at some intermediate point on the scale of perfection. The German naturalist Lorenz Oken even saw fit to describe 'lower' animals as 'human abortions', and the French naturalist Jean-Baptiste Robinet described them as 'the apprenticeship of nature in learning to make man' (Gould 1977, p. 36f.).

The idea that there is a parallel between the two series became a serious, explicit theory in the late 18th and early 19th centuries. There were variations in the specific beliefs of the different thinkers, of course, but the basic idea was common to them all: that there is a close parallel between the stages of development of the human being and the 'scala natura'. The human being, in its development, thus passes through stages corresponding to 'lower' organisms. Importantly, this needn't entail that *every single* organism in the natural world is *very accurately* represented in the development of the human being. Instead the 'essential features' of the different *classes* of lower organisms will be represented.[5]

With the caveats included, can such a theory be tested? Gould (1977, p. 6) writes: 'Recapitulation was largely impervious to empirical disproof by accumulated exceptions.' However, it was only 'impervious to empirical disproof' in the same sense that any theory is: one can always preserve the 'hard core' of a theory if one really wants to, adjusting auxiliaries or background assumptions as necessary. This is compatible with the theory being testable, in the sense that it might become increasingly unreasonable to keep adjusting the theory to wriggle out of uncomfortable empirical facts. In the case of the recapitulation theory, the claim that the human being passes through certain stages corresponding to 'lower' classes of organisms can certainly be investigated. An obvious animal class is the class of fish, and an obvious candidate for an essential feature of a fish is its gills. This is reasonably considered an *essential* feature of fish, since all fish have gills, and no fish can function without them. Thus one reaches a prediction of the theory of recapitulation: the human being, at some stage in its development, would be expected to include gill-like features. And it was obvious to those at the time—given the empirical knowledge of the day—that this must happen very early on in the development of the human embryo.[6] Thus it was only a matter of time before

[5] Gould (1977, Chapter 3): 'C. G. Carus, an influential Naturphilosoph and supporter of Oken, wrote: "Each degree of development of a superior animal constantly recalls a determined form of an inferior organism; but between the two there is no complete identity, but only a resemblance of fundamental nature or essence" (1835, 2: 438).' Such caution concerning the exactness of the parallel was already clear at the beginning of the 19th century. For example, in 1808 Tiedemann wrote: 'Every animal, before reaching its full development, passes through the stage of organisation of one or more classes lower in the scale.' And in 1809–11 Oken wrote: 'During its development the animal passes through all stages of the animal kingdom. The foetus is *a representation* of all animal *classes* in time' (emphases added; see Russell 1916, Chapter 7). Meckel actually wrote in his 1811 essay: 'I am indeed very confident that in our recent knowledge of the embryo of the higher animals no complete equation of the same with the lower ones can be made; but [...] some not uninteresting moments can be established...' (p. 2, my translation).

[6] Studies of the development of the human embryo/foetus were quite numerous at the time of Meckel's prediction, in 1811. Many examples can be found in Adelmann (1966). Harvey studied the human foetus in the mid-17th century (see Adelmann, pp. 1609, 1761). Wrisberg had looked at a 10-week

one of the recapitulationists stated explicitly that gill slits ought to be found in the early human embryo.

JF Meckel was a major figure in embryology at the beginning of the 19th century; he was known at the time as 'the German Cuvier'. And his work of 1811 (Meckel 1811) is, in the words of Gould (1977, Chapter 3), 'Naturphilosophie's major statement of recapitulation'.[7] It is here that the gill slit prediction is found. On page 25, after discussing a specific period of embryological development, he writes:

> Vielleicht findet sich sogar eine sehr frühe Periode, wo der Embryo der höhern Thiere auch mit inner Kiemen versehen ist.
>
> Perhaps there is even a much earlier period when the embryos of the higher animals are also furnished with inner gills.
>
> (Meckel 1811, p. 25, my translation)

And sure enough, some 15 years later, gill slits were found. In 1825 the German embryologist Heinrich Rathke published on his observations of gill slits in the embryos of birds and mammals, comparing the slits with those of a dogfish shark. Others immediately followed up this line of research, including the German scientist Karl Ernst von Baer. Rathke and von Baer discovered gill slits in the *human* embryo around 1827–8.[8]

On the face of it we have here an example which provides evidence against the claim that novel predictive success is strongly correlated with (approximate) truth. Meckel's prediction concerned a novel phenomenon, in the sense

old human embryo in 1764, and Rosenmüller had looked at a nine-week old human embryo in 1802 (p. 1775). Heuermann had looked at a five-month old human foetus in 1755 (p. 1784). Meckel had himself examined 'a considerable number of human embryos' at the start of the 19th century (p. 1792).

[7] For the movement sometimes referred to as 'Naturphilosophie', see Gould (1977, p. 35ff.). It is worth noting that some form of recapitulation goes back at least to the ancient Greeks (see Gould 1977, Chapter 2), though there is good sense in focusing on the later 18th and early 19th centuries.

[8] Rathke (1825a; 1825b). Von Baer (1827, p. 557) writes: 'Ich zweifle indessen nicht, dass im Menschen und vielleicht in allen Land-Wirbeltieren ursprünglich vier Kiemenspalten sind' ('I do not doubt, however, that there are originally four gill slits in humans and perhaps all terrestrial vertebrates') (my translation). Menz (2000, p. 165) writes: 'In 1828, Rathke, with "the help of a good magnifying glass" and "in bright sunshine", also found gill slits in a human embryo from the sixth or seventh week of pregnancy. He formulated this finding more cautiously [than previous publications]; in the title [of his 1828 publication] he spoke only of "gill indications"' (my translation). See also Valentin (1835), p. 487: 'Von Baer, who had provided a magnificent drawing of gills in young embryos of the dog, found (1827) no gill slits on the smallest embryos of all higher vertebrates; in humans, however, he saw them most clearly in a six-week-old embryo, where, however, they had already begun their retrograde metamorphosis' (my translation).

that gill slits in 'higher' animals were not known in 1811. In addition the prediction was quite specific, and thus 'risky' in the sense that the predicted phenomenon could certainly be investigated, and it might have turned out that the prediction failed—there might have been no gill slits at any stage of any mammalian embryo. This would have been damaging for the theory. Thus this presents itself as the kind of prediction which can carry significant evidential weight. At the time—with the theory of evolution unconceived—it would also have seemed to meet with Mayo's 'severity criterion' (Mayo 1991, p. 529) which requires that there be a very low probability that the prediction be successful if the theory is false.[9]

What if—borrowing from the scientific realist[10]—we employ the 'divide and rule' strategy? We may ask: Did Meckel's recapitulation theory have this success because it includes within it certain 'working' hypotheses which are indeed approximately true, with only 'idle' parts of the theory being radically false? Realists over the years have shown significant ingenuity when it comes to the employment of this strategy—what seems to be 'working' has sometimes been claimed by the realist to somehow be 'idle'. However, in the case of recapitulation theory it is hard to find *any* claims we might now describe as 'approximately true'. The idea of a linear scale of organic perfection is radically false. So too, the idea of a single biological 'force' or 'power', with more or less strength in different organisms, is radically false.

What about the core recapitulationist claim that the human embryo passes through the adult forms of the 'lower' classes of animals? This is clearly false, but it might be considered to contain a thread of truth in the form of the following proposition (*):

> (*) The human embryo temporarily takes on certain features during its development which are found in the *adult* forms of other creatures.

This proposition is 'contained within' the statement of recapitulation in the sense that if one believes in recapitulation one is thereby also committed to (*).

[9] Note that Mayo (1991) should really have said 'if the *working parts* of the theory are false', since a false theory may be false merely because certain 'idle wheels'—doing no empirical work—within that theory are false.

[10] As discussed in Chapter 2, this book certainly isn't a defence of 'realism' as it is usually construed within the context of the scientific realism debate. But some of the realist's defensive strategies may be useful. Just as the realist wants to say of past, now rejected scientific ideas 'I wouldn't have been a realist about those ideas', so too I want to say that I wouldn't have labelled those ideas future-proof. Reasons for *not* being a 'realist' about various scientific ideas can also stand as reasons for *not* labelling those ideas 'future-proof'.

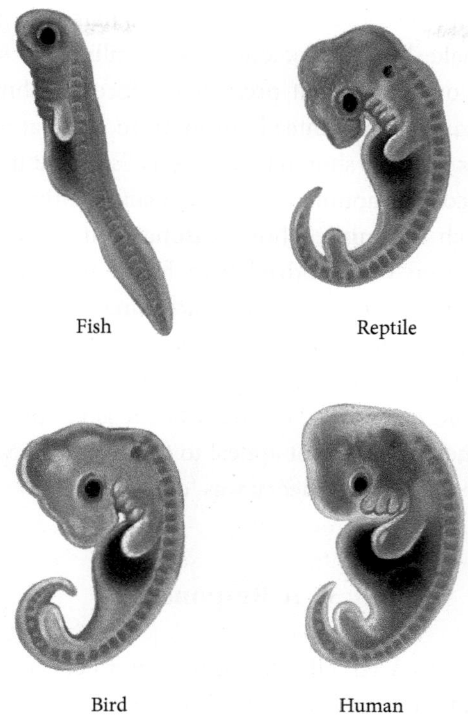

Figure 3.1 The human embryo at five weeks (approx.) compared with other vertebrate embryos

However, this is no longer a theoretical statement: it had been established empirically long before Meckel. For example, the human embryo temporarily has a tail (Figure 3.1), and this is relatively easy to observe.[11]

[11] The precise date of the discovery of the temporary tail in the human embryo is difficult to establish, but it was certainly known by Meckel's time. It is observable to the naked eye, so observation of the human embryo's tail did not require developments in microscopy (the human embryo is the size of a lentil at five weeks). Wolff speaks of the 'tail' of the early chick embryo frequently; e.g. he wrote in 1764: 'We know that subsequently the embryo finally draws together to such an extent that its stout head almost touches its tail' (quoted in Adelmann 1966, p. 1770). Meckel knew Wolff's work very well; in 1812 he translated into German Wolff's 1768/9 classic work *De Formatione Intestinorum* on the formation of the intestine (in the chick). Charles Bonnet also mentions the 'tail' of the early chick embryo in 1769, writing: 'When the chick begins to become visible, it appears in a form very similar to that of a very small worm. Its head is large, and to this head is attached a sort of tapering appendage. It is, however, in this appendage, so similar to the tail of a small worm, that the trunk and limbs of the animal are contained' (quoted in Gould 1977, p. 25). The tail of the human embryo would have been accepted long before it was observed, on the basis of an inference from the fact that all non-human vertebrate embryos have a temporary tail, and all vertebrate embryos look the same at an early stage of development. Certainly Meckel and others would have seen Sömmerring's drawings of the early human embryo found in Sömmerring (1799) (Plate 1, Figure 2 shows the tail of the human embryo clearly).

In addition, it isn't clear that (*) is sufficient to make the gill slit prediction. Meckel did not make lots of predictions willy-nilly, in a scattergun style.[12] Instead he picked out the gill slit prediction, since (I submit) he felt it was probably true (even if not guaranteed) given his recapitulation hypothesis. But it is far less obvious that one should make Meckel's gill slit prediction from (*). In weakening the core recapitulationist claim sufficiently to reach something true, we have much diminished how reasonable it is to make the gill slit prediction. Proposition (*) is indeed true, but it is compatible with many thousands of speculative 'predictions' concerning what might be found in the human embryo. So the truth of (*) does not explain the fact that Meckel hit upon a single, true prediction, using his theory. Certainly (*) follows from Meckel's theory, and certainly (*) is true, but it isn't sufficient to make the prediction. Thus the realist cannot appeal to (*) to explain Meckel's success in terms of truth; Meckel's wider theory was 'doing work'.

3. A Response?

Meckel's suggestion vis-à-vis gill slits is often described as a 'prediction' in the literature. One example from the time is Valentin (1835, p. 485) who uses the German word 'voraussagte'. Three modern examples are: Gould (1977, p. 46), Stanford (2009, p. 383f.), and Menz (2000, p. 149). The latter straightforwardly states: 'Meckel (1781–1833) already in 1811 anticipated and predicted the occurrence of gill slits in embryogenesis' (my translation). However, it may be noted straight away that Meckel uses the word 'vielleicht' ('perhaps'/ 'maybe'). The question may be raised: is Meckel predicting, or is he merely speculating? Certainly in using the word 'vielleicht' he leaves open the possibility that gill slits will *not* be found. This protects the theory of recapitulation from falsification (serious disconfirmation), but at the same time it seems to diminish the measure of confirmation of the theory when the prediction *is* confirmed.

However, as noted already, Meckel does pick out this one specific phenomenon, and commits it to print. Given that nothing like gill slits had ever been (knowingly) observed in 'higher' embryological development at that time, it would have been rather bold for him to assert very strongly that gill slits *would*

[12] Cf. Vickers (2019, pp. 580–1) for more detailed discussion of 'scattergun-style' novel predictive success in the case of Velikovsky's theory of the development of the solar system.

be found, especially given that he hadn't seen them himself, in his own numerous examinations of human embryos and foetuses.[13]

In addition, Meckel's psychology is not relevant here; we needn't agonise over why he chose to say 'perhaps' instead of 'probably' or something stronger. The most important thing, for the purposes of weighing the degree of confirmation, is the tightness of logical fit between the theoretical ideas and the predicted phenomenon. And the fit is very tight indeed, since (i) fish are a very prominent class of organism, (ii) all fish have gills, and (iii) gills are functionally essential for fish. If there really were a parallel (of essential features) between the ontogenetic development of the 'higher' embryos and the series of perfection of the adult 'lower' classes of organism, then gill slits were to be strongly expected somewhere in the development of the human embryo. Meckel probably saw that, but chose to proceed with caution by using the word 'vielleicht'.

Let us accept, then, that the theory of recapitulation genuinely predicted gill slits in the human embryo. Does it follow that the theory of recapitulation was strongly confirmed by the success of this prediction? Certainly there were some who were impressed with this kind of evidence for recapitulation. Rathke wrote in 1825 of 'a lovely confirmation' of recapitulation in the case of the development of kidneys in the 'higher' animals when compared with the permanent form of kidneys in 'lower' animals.[14] The following year, comparing the development of birds, newts, and chicks, and with explicit reference to gill slits in the embryo of the chick, Rathke wrote of 'a most important and obvious piece of evidence ['einen höchst wichtigen und augenfälligen Beleg']

[13] In fact, the gill slits just *might* have been seen before 1811; possibly Malpighi and/or Wolff had seen them, but had quickly dismissed them (Valentin 1835, p. 485f.). Further evidence that they may have been seen before 1811 is provided in Volume III of Adelmann (1966), p. 1514. On the basis of this evidence, it may even be speculated that Meckel had already seen what he was predicting. However, I find it highly unlikely that Meckel made a conscious decision to *pretend* not to know about the gill slits when he predicted their existence in his 1811 essay; it is much more likely that he really hadn't seen them, or else he *had* 'seen' them (loosely speaking) but without realising what he was looking at, a phenomenon that is common enough in science.

[14] I take the words 'a lovely confirmation' from the original German: 'eine schöne Bestätigung' (Rathke 1825c). Rathke goes on to compare this development of the kidneys with the way gills give way to lungs when tadpoles develop into frogs: 'Here, moreover, the course of development is similar to that which we noticed in the respiratory apparatuses of Batrachia and birds. In other words, since these repeatedly mentioned bodies completely disappear in some individuals while the kidneys develop further and further, we see a relationship and a behavior similar to what we observe in the gills and lungs of Batrachia. The one structure, which indeed can also be compared with a similar structure in animals lower in the scale, gives way to another which is later actually to take over its function materially.' (Here I use the translation found in Adelmann 1966, p. 1826.)

for the assertion that the higher vertebrates absorb in their formation the development of the lower ones'.[15]

However, we must distinguish sharply between the personal impression of Rathke and others of the degree of confirmation, and the actual degree of confirmation. In general, it is well known that what may *seem* like a case of strong confirmation may not be as strong as it appears. This can even occur in contemporary science: one need only look at the literature surrounding the recent 'replication crisis' to find numerous concrete examples of a gap between the *apparent* strong confirmation of a hypothesis, and the actual confirmation of that hypothesis. These examples even involve rigorous contemporary statistical methods, and can appear very convincing indeed, not only to the layperson.

When we ask what factors are relevant to confirmation, one obvious place to look is the Bayesian equation: $p(T,E) = p(T)p(E,T)/p(E)$. Even if we are not keen on 'Bayesianism', some clear rules of thumb drop out of this equation. We have already discussed one of these: if the probability of *the predicted phenomenon given the theory*, $p(E,T)$, is higher, then the probability of *the theory given the evidence* (the posterior $p(T,E)$) is higher. To put this informally, if your theory *entails* something, then that is more significant for the confirmation of the theory than if it merely *suggests* something: in the former case the prediction is riskier. In our case, recapitulation theory doesn't *entail* gill slits in the human embryo, but I have argued that the probability is high, so the lack of entailment doesn't affect the degree of confirmation much.

Another important rule of thumb is this: if the probability of the predicted phenomenon is *already* high, for some reason (independent of the fact that the theory predicts it), then the measure of confirmation is reduced. In other words, if a theory predicts something that everybody already expects to be the case, and then (behold!) the phenomenon is in fact observed, that 'predictive success' shouldn't increase our belief in the theory very much. To give an example: the theory of evolution (together with a few basic background assumptions) predicts that new fossils are highly likely to be discovered over the next five years. But when they *are* discovered, that arguably does nothing to confirm the theory, since anyone who doesn't believe the theory will *also* expect new fossils to be discovered. One can predict that new fossils will be discovered at some point in the next five years simply on the basis of past

[15] My translation. The passage can be found in Menz (2000, p. 159), and comes from an article Rathke wrote in 1826, but which appeared in 1828, entitled 'Ueber die Entwickelung der Athemwerkzeuge bei den Vögeln und Säugethieren' ['On the development of the respiratory system of birds and mammals'].

experience. That is, one can make this prediction *even if one doesn't hold a theory about the origin of fossils*. This thought—that one can sometimes make a prediction on the basis of one's past experience, or background empirical knowledge, regardless of what theory (if any) one holds—will be important in what follows.[16]

4. Von Baer

When Meckel made his prediction in 1811 it seemed—to the uninitiated—to be extremely bold. The *uninitiated* would have put the prior probability of gill slits in the human embryo, p(E), very low. And if p(E) is really very low, then the posterior p(T,E) is very high. In other words, the successful prediction confirms the theory very strongly, meaning that many of those previously disposed to be sceptical should be won over. But somehow (*pace* Churchill 1991, p. 11) this did not actually happen when Meckel's prediction was confirmed in 1825–8: there is no evidence[17] that sceptics of recapitulation were won over by Meckel's successful prediction.

One important reason for this has to be Von Baer's influential attack on the theory of recapitulation. This attack would have been well known in the community at least as early as 1826.[18] In a paper dating to 1826 Von Baer criticises recapitulation as follows:

> In the first place we safely base the conviction, that all the forms of animals are by no means to be considered as uniserial developments from the lowest to the highest grade of perfection. [...] It has been concluded by a bold generalization from a few analogies, that the higher animals run in the course of their development through the lower animal grades. [...] We hold this to be not only untrue, but also impossible. [...] we observe that animals in the course of their development are more or less similar only to lower stages of the same type. (Von Baer 1826, p. 184)

[16] Cf. Dorling (1992, p. 368), attempting to judge the degree of confirmation afforded to (Daltonian) atomism in light of the truth of the laws of constant and multiple proportions. There, as here, one has to judge how probable the phenomenon in question is 'merely on the basis of the direct observational evidence so far available'.

[17] Churchill (1991, p. 11) writes that in 1828, 'The law [of recapitulation] had received a significant boost from Rathke's recent [1825] discovery of gill slits in avian and mammalian embryos.' But I have searched quite thoroughly and found no evidence that anybody changed their mind in light of this evidence. Churchill's term 'significant boost' is (probably) not warranted.

[18] Von Baer (1828, p. 190) notes that he expressed his doubts concerning the 'scala natura' in 1823. On this, see Russell (1916), p. 121.

This attack involved not only (i) criticisms of the theory of recapitulation, but also (ii) an alternative account for why we should expect to find certain striking comparisons between different developmental stages of different organisms. These two things combined served to disconfirm the theory of recapitulation, meaning that even if our degree of belief was *poised* to increase considerably on account of the success of the gill slit prediction, it would be held back on account of other considerations.

First of all, the theory seems to predict phenomena that are *not* found in nature. Von Baer (1828, pp. 192–3) provides several examples of phenomena going against the thesis of recapitulation. As Russell (1916, p. 121) writes, 'Von Baer had soon found in the course of his embryological studies that the facts did not at all fit in with the doctrine of parallelism.' For example, Von Baer (1828, p. 193) writes: 'If the law we are engaged in investigating [recapitulation] were correct, no conditions which are permanent only in the higher animals could be a transitory stage in the development of particular lower forms. But a great number of such conditions are demonstrable.' He then 'presents a list of features that are fixed in adult mammals, but transitory in bird embryos' (Gould 1977, p. 54; see also Gould 1977, p. 53). Von Baer (1828) notes that 'the chick embryo, at one stage, has a heart and circulation very much like that of a fish, but at the same time it lacks "a thousand other things" that all adult fishes possess.'[19]

Secondly, Von Baer had his own theory of embryological development, which was *also* consistent with the empirical evidence Meckel and others had put forward as evidence for recapitulation. Von Baer claimed that there was a basic 'archetype' or 'fundamental form' for each class of animal.[20] All vertebrates start out in this fundamental form; indeed, at an early stage of embryological development it is impossible to tell apart the embryos of birds, lizards, mammals, and so on. Von Baer wrote,

> The embryos of Mammalia, of Birds, Lizards and Snakes, probably also of Chelonia, are in their earliest states exceedingly like one another, both as a whole and in the mode of development of their parts; so much so, in fact, that we can often distinguish the embryos only by their size. In my possession are

[19] Von Baer also notes that, 'Parts that characterize higher groups should appear late in embryology, but often do not. The vertebral column of the chick, for example, develops very early' (Gould 1977, p. 55).

[20] See e.g. Von Baer (1828, p. 210): '[T]he feet of Lizards and Mammals, the wings and feet of Birds, no less than the hands and feet of Man, all arise from the same fundamental form.'

> two little embryos in spirit, whose names I have omitted to attach, and at present I am quite unable to say to what class they belong. (1828, p. 210)[21]

As the embryos of different organisms develop they do so in different ways, such that one can increasingly see the differences between them. But some organisms develop further from the 'archetype' than others. Von Baer is very clear on this crucial point: 'It is only because the least developed forms of animals are but little removed from the embryonic condition, that they retain a certain similarity to the embryos of higher forms of animals' (1828, p. 214). When it comes to gill slits, all vertebrates have them at an early stage, because these features are part of the fundamental form of the vertebrate. Fish develop only a little from the fundamental form, and thus retain their gill slits, but mammals develop much further such that their gill slits completely disappear. As Von Baer writes, 'Fishes are less distant from the fundamental type than Mammalia, and especially than Man with his great brain. [. . .] [T]he embryo of Man is unquestionably nearer to the [adult] Fish than conversely, since he diverges further from the fundamental type' (1828, p. 220).

Appreciating the similarities and differences between recapitulationism and Von Baer's archetype theory is not straight forward. Consider Ospovat (1976, p. 8), commenting on Von Baer's archetype theory:

> The result is a certain resemblance between, for instance, the adult fish, which stands close to the archetype, and the mammalian embryo, which, at an early stage, is also but little removed from the archetype of the vertebrates. It is because such resemblances are a necessary consequence of his own theory that Von Baer remarked that the difference between his view and the theory of recapitulation was not as great as it at first sight appeared. And it is for this reason that the historian must be careful not to suppose that every biologist who in the decades after 1828 spoke of resemblances or analogies between embryos and adults was an adherent of the theory of recapitulation. Some, perhaps the majority, were advocates of the theory of diverging development first proposed by Von Baer.

To demonstrate the issue, consider this statement from Meckel (1811): 'Both the arrangement of the vascular system in general, and of the heart in

[21] In his 1811 essay (p. 5) Meckel had written that at first there is 'a perfect equality' ('eine vollkommne Gleichheit') between the 'highest animal' and 'several reptiles and fish'. So on this point there is certainly no disagreement between Von Baer and Meckel.

particular, offers the most interesting agreement between the embryo of the higher animals and the perfected [fully formed, adult] lower animals' (p. 8, my translation). This sounds like recapitulation, and Meckel put it forward as part of his case for recapitulation, but in fact Von Baer would have found little to disagree with here. He would only have disagreed on *why* there is this agreement—for Von Baer it doesn't have to do with recapitulation; it instead has to do with the fact that the 'lower' animals do not move far in their development from the fundamental type.

With a different theory on the table which could also account for the gill slits in the human embryo, the confirmation of Meckel's theory of recapitulation is significantly reduced. This is easily seen if we return to Bayes' equation, and we expand p(E) in the denominator as follows:

$$p(T,E) = \frac{p(E,T)p(T)}{p(E)} = \frac{p(E,T)p(T)}{p(E,T)p(T) + p(E,\neg T)p(\neg T)}$$

For a significant confirmation, we need p(E) to be quite low. This is most obviously achieved when the predicted phenomenon, whilst expected on the theory under question, is definitely *not* expected otherwise; in other words *the probability of the phenomenon if the theory is false*, p(E,¬T), is low. But with Von Baer's archetype theory on the table, this term increases significantly. It turns out that one doesn't need recapitulation theory to make sense of the mammalian gill slits.[22]

So it happens that at the same time Meckel's gill slit prediction was confirmed by observation, and could potentially have served to significantly confirm the theory of recapitulation, Von Baer countered with both (i) awkward empirical facts, in tension with Meckel's theory, and (ii) an alternative theory of the relevant phenomena. This goes some way towards explaining why there is no textual evidence of sceptics vis-à-vis recapitulation being won over by the success of Meckel's prediction. The evidential situation didn't necessarily favour Von Baer's theory right away, but there was certainly sufficient doubt to put the brakes on recapitulation theory. It also explains why Meckel was a bit lukewarm when in 1833 he wrote:

[22] There had been other 'archetype theories', long before Von Baer (1828). For example, Goethe had an 'archetype theory' in the 1780s, which was inspired by others coming before him such as Buffon, Herder, and Camper (see e.g. Russell 1916, Chapter 4; Wells 1967a, 1967b). But Von Baer's theory was altogether more scientifically serious, and Von Baer in 1828 made it clear how his theory accounted for the phenomena commonly put forward as evidence for recapitulation (including the phenomena cited in Meckel's 1811 essay).

> The discovery of them [gill slits] is indisputably highly interesting because it provides a new piece of evidence [Beleg] for the correctness of the law of development [recapitulation]. In spite of the fact that at first one might have doubts about the meaning of the lateral divisions, and I myself, in spite of my well-founded predilection for that law [recapitulation], had such doubts at first [...], surely now, after so many good anatomists have confirmed the discovery, in such different animals, the correctness of their interpretation has been proven. (From Menz 2000, p. 168, my translation)

He would have loved to have been able to say, 'The gill slits in the human embryo only make sense if we accept recapitulation.' He could not say that in 1833.

5. The Argument from Empirical Knowledge

We may ask again to what extent, if at all, a 'no miracles' argument can be applied to Meckel's predictive success. It might well be suggested that the emergence of Von Baer's theory in 1826–8 does nothing to help us understand how Meckel could have predicted gill slits in 1811 by use of a radically false theory. It is also possible that Von Baer's theory might have come along later, and Meckel's gill slit prediction might have been confirmed as successful years earlier (e.g. 1812). In that situation, we may ask, would it have been rational to believe recapitulation theory, on 'no miracles' grounds? If so, then the realist could still be in trouble here, since she needs realism to be compatible with alternative possible worlds where science developed in a slightly different way, for example because results emerged in a different order (cf. Section 5 of Vickers 2018a). The empirical disconfirmations of recapitulation theory put forward by Von Baer in 1828 might also have been articulated much later, long after the gill slit confirmation. Is it only by a fluke of the particular (surely contingent) historical development of science in the years 1800–30 that the evidence was only partially misleading, and not *highly* misleading?

In Section 4 it was noted that the existence of Von Baer's archetype theory serves to increase p(E) in the Bayesian equation. This in turn reduces p(T,E), which is what the realist wants—the realist does *not* want to have to accept that she would have believed Meckel's theory at the time, and then changed her mind at a later time. As discussed in Chapter 2, realism is a cautious thesis, designed to tell us the circumstances in which we can be *very* confident that the evidence is sufficient to make a knowledge claim. Needless to say, the realist's position starts to look silly if there are examples where she would have said 'We now *know* this—this is future-proof', then later changed her mind.

Suppose Von Baer's theory had not been available; would p(E) have then remained rather *low*, suggesting that one's degree of belief in the theory should become very high? Perhaps not: there are other good reasons—other than the existence of an alternative theory—for putting a reasonably high value on the prior probability of the gill slit phenomenon p(E). Sometimes the empirical knowledge at one's disposal already suggests other phenomena, even in the absence of any relevant theory. And in the case of early embryology and comparative anatomy during the 17th, 18th, and early 19th centuries, inferences from known phenomena to unknown phenomena were extremely widespread.[23]

One thing that was absolutely beyond doubt in the 17th and 18th centuries was that all vertebrates are similar in many respects, at least for a period. As Wells (1967b, p. 537) writes:

> When Goethe began to study anatomy in 1781 it was well known that vertebrates have a great deal in common, that for instance a bird or a horse has much the same bones as a man [...]. Well before Goethe it was even realized that the similarities (not restricted to the skeleton) are so marked that structures which in animals serve an obvious function are present in man even when useless to him and that in such cases the common type underlying vertebrate forms is a more fundamental factor than utility.

'Predictions' concerning what would probably exist, or what *might* exist, in one type of vertebrate, given that it existed in another, had been commonplace for centuries.[24] One of the most dramatic such predictions concerned the existence of the mammalian egg. Already in the 1630s William Harvey spent two months conducting a thorough search for the mammalian egg in pregnant deer.[25] The time and energy devoted to this search testifies to the strength of

[23] Such empirical knowledge goes back at least as far as Aristotle. In his *Historia Animalium* Aristotle described in detail the anatomy of more than 500 types of animal, including less obvious examples such as cuttlefish, oysters, sponges, and chameleons. He also studied the embryological development of the chick. For discussion, see e.g. Russell (1916, p. 2ff.). Adelmann (1966, Volume 2) covers the early history of embryology in greater detail.

[24] Such considerations brought Goethe, in 1795, to (correctly) infer the existence of the intermaxillary/premaxilla bone in (early) human development, on the basis that it had been observed in other vertebrates (see Wells 1967a). However, the status of Goethe's prediction was still in doubt in the early 19th century, so its significance as a source of inspiration for Meckel is limited.

[25] Meckel (1811) explicitly draws on Harvey's work on pp. 7–8. He references Harvey's *De Motu Cordis*. He also quotes Harvey on p. 8.

Harvey's conviction that such an egg was in fact there to be found. And this is just another way of saying that Harvey *predicted* that the mammalian egg existed. Indeed, this is just as much an example of 'novel predictive success' as Meckel's prediction of gill slits in the human embryo—there *is indeed* a mammalian ovum, and evidence for its existence gradually came in during the 17th and 18th centuries, with Von Baer finally observing it directly in 1827.

Such examples would have been well known to anatomists and embryologists in the 18th and early 19th centuries. At the time of Meckel's essay of 1811 the mammalian ovum was yet to be observed directly, but by then the evidence for its existence was overwhelming. Here, then, was a clear example of how very different vertebrates share features that you might not initially expect. Chickens and fish clearly lay eggs to reproduce, but it turns out so too do human beings. Already in 1667 in his *A Model of Elements of Myology*, Nicolas Steno had shown that the mammalian ovary—previously referred to as the 'female testicle'—is analogous to the ovary of the dogfish shark. As such knowledge of surprising comparisons between very different types of vertebrate increased, it became more and more reasonable to ask what *other* similarities, previously unsuspected, might exist. One did not need to believe in recapitulation theory to make these comparisons, and wonder about other possible similarities. Even those who doubted the more radical thesis that there were serious comparisons to be made between vertebrates and non-vertebrates thought that comparisons between the different vertebrates were reasonable. And this, of course, included comparisons between human beings and fish.

The most numerous examples of interesting and surprising comparisons between vertebrates concerned those between different vertebrate *adults*. But there were also known comparisons between embryonic 'higher' mammals (including humans) and the adult forms of 'lower' vertebrates. A good example is provided in Meckel (1811).[26] It was well known at that time that when the human heart is four weeks old it goes through a stage where it very closely resembles the heart of an adult fish (Figure 3.2). At around six or seven weeks it resembles the heart of an adult reptile (see Gould 1977, p. 47). These comparisons could not be denied, whether one held to the theory of recapitulation or not.

[26] Meckel's essay of 1811 includes many possible comparisons, put forward in support of recapitulation. Some are quite reasonable, whereas others are highly speculative. Meckel was known for being 'imaginative' with his analogies between different types of organism. Lenoir (1989, p. 59), discussing Meckel, refers to 'unwarranted inferences based on a plentiful and imaginative use of analogical reasoning'.

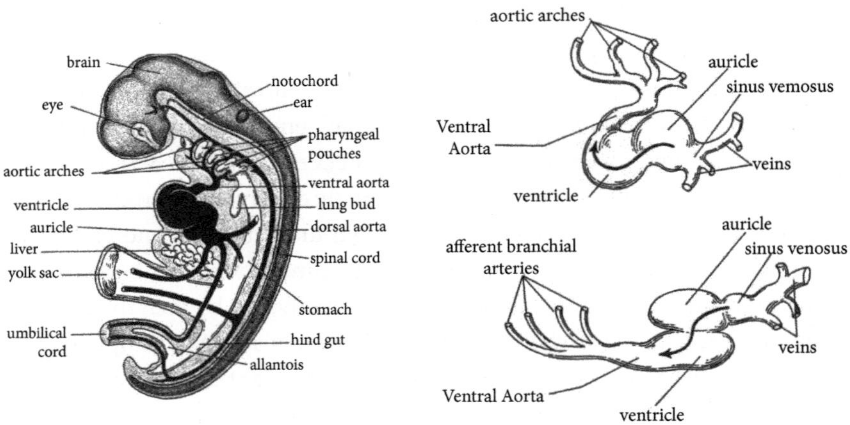

Figure 3.2 Comparison of the human heart when the embryo is four weeks old with the heart of an adult fish

In addition, zoologists of the day knew about *another* case of a vertebrate starting out at an early stage of its development with gill slits, and later developing—and breathing with—lungs. This is the case of the frog. Tiedemann (1808) had written about this case, quite possibly influencing Meckel's prediction of gill slits three years later. Certainly Meckel had read Tiedemann's work; in his 1811 essay he references Tiedemann's 1809 paper 'Anatomie des Fischherzens' (Meckel 1811, p. 11).

First, Tiedemann states (a version of) the theory of recapitulation: 'Every animal, before reaching its full development, passes through the stage of organisation of one or more classes lower in the scale' (quoted in Russell 1916, p. 91). He then gives the frog as a clear example: it starts out like a worm, then it develops gills and becomes like a fish (when it is a tadpole), before finally losing the gills to breathe with lungs. Not everybody agreed with Tiedemann's recapitulation interpretation of the development of the frog. But what *would* have been obvious to everyone was the more basic empirical fact: that in the transition from tadpole to frog we have a clear example of a vertebrate starting out with gills, and later developing lungs. And given all the previous examples of surprising, hidden similarities between all different types of vertebrate, one could certainly be forgiven for wondering whether the transition from gills to lungs might occur in *other* vertebrates, not only frogs and toads. In other words, it increases the prior probability p(E) of gill slits in the mammalian embryo. And this, in turn, *de*creases the confirmation of recapitulation theory when mammalian gill slits are indeed observed. Meckel (1811, p. 25), when he makes his prediction, explicitly makes a comparison between the known development of Batrachia (frogs, salamanders), and the possible development of 'higher mammals'. The early gills in humans might be 'replaced' ('ersetzt') with some other feature during the development of the higher mammals, just as is seen in frogs. Numerous examples of such 'replacement' were known to the embryologists of the early 19th century.

Regarding some of these examples there is a difficult question: To what extent was empirical knowledge driving the prediction, and to what extent was theory driving it? Recapitulationists such as Tiedemann and Meckel would have clearly interpreted the gill slit result as intimately connected with recapitulation theory. However, if background empirical knowledge raises the prior probability of human gill slits high *enough*, then it is reasonable to claim that the theory is really doing little or no work. This stands, I submit, *even if* a theorist derives the prediction *from* the theory in question. How so? Well, if a wealth of phenomena *W* already suggests another phenomenon *E*, then if one develops a theory to account for *W* it will be no surprise if the theory also suggests *E*. Thus one reaches *E* from the theory, perhaps, but previously the theory was reached from *W*. So although it may seem as if the theory is doing the work, the theory can only do that work because it was built on *W* in the first place. The wealth of phenomena *W* is still doing much of the work to generate the prediction of *E*, but now *via* the theory, rather than directly (Figure 3.3).

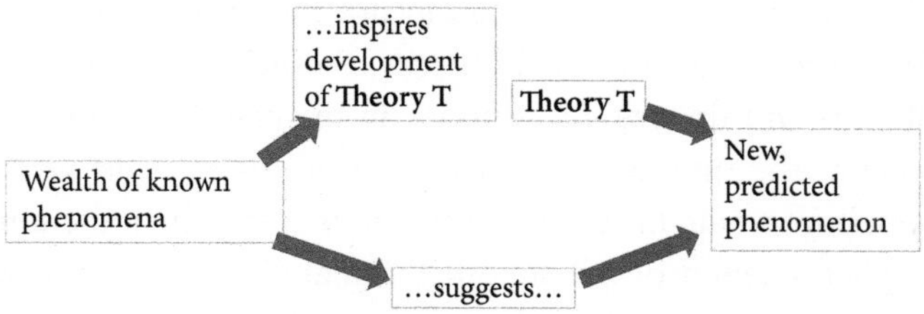

Figure 3.3 Two routes—top path and bottom path—from a wealth of known phenomena to a new, predicted phenomenon; the more secure the inference on the bottom path, the higher the prior probability of the new phenomenon, and thus the lower the confirmation of **Theory** T when the prediction is confirmed.

It should be noted, too, that Meckel knew the literature (and therefore the relevant 'background' empirical knowledge) very well. In his papers he references all the key players in the history of comparative anatomy and embryology. Specifically, in his 1811 essay he references, among others, William Harvey, Marcello Malpighi, Albrecht von Haller, Friedrich Tiedemann,

Caspar Friedrich Wolff, Georges Cuvier, William Cheselden, Walter Needham, Jan Swammerdam, Johann HF von Autenrieth, Samuel Thomas von Sömmerring, Alexander Monro (secundus), Georg Wilhelm Steller, and Félix Vicq d'Azyr, citing their prior work on the anatomy and embryological development of organisms including the manatee, seal, turtle, crocodile, chicken, squirrel, wolf, stag, pig, fox, cat, conger eel, guinea pig, hare, lynx, porcupine, elephant, platypus, camel, otter, beaver, bottlenose dolphin, lizards, salamanders, frogs, snakes, apes, all different types of fish, and different types of rat and mouse. An extraordinary number of interesting parallels can be drawn when one considers the way in which all of these different organisms develop, and one compares them with the development of the human being. Of course, Meckel draws on these relationships between different organisms to support recapitulation, but the empirical work he cites actually supports the specific prediction he makes, even for somebody *not* inclined towards Meckel's recapitulation theory (cf. Figure 3.3).

This line of thought cannot be pushed too far. Sometimes the empirical base used to construct the theory does *not* (strongly) suggest the predicted phenomenon. The Poisson-Arago 'white spot' is surely a case in point. Indeed, in the case of the white spot, Poisson was quite sure that the phenomenon would *not* exist, and he wished to test the prediction with the goal of falsifying (seriously disconfirming) Fresnel's theory. Poisson's confidence that the predicted phenomenon would not exist testifies to the fact that phenomena associated with light up to that time did not indicate that one might find a white spot, as predicted by Fresnel's theory. Fresnel's theory was surely doing significant work in this case. By contrast, in the case of the prediction of gill slits, nobody at the time *who was aware of all the relevant empirical background knowledge* would have been confident that the phenomenon would not exist. The phenomenon of gill slits in the mammalian embryo was already suggested by prior empirical results, in a way the Poisson-Arago white spot was not. This is a crucial disanalogy demanding a different attitude towards the evidential significance of the successful prediction in the two cases.

6. Conclusion

Let's back up a bit and ask: Why was this case supposed to be significant? The worry was that it is an example of strong evidence (novel predictive success) for a radically false theoretical idea. And if many such examples could be identified in the history of science, that might speak significantly against the idea that 'evidence' and 'truth' are strongly correlated. This would undermine the claim that the 'strong evidence' for many contemporary scientific theories indicates that they are future-proof.

When we look at the finer details of the case, though, it is clear that it has no power to speak against contemporary scientific theories. The success of the prediction didn't even have a big impact at the time,[27] since by the late 1820s Von Baer had argued quite decisively against recapitulation, and in favour of an alternative account of the relevant phenomena (including the phenomenon of gill slits in the human embryo). But even if Von Baer had never existed, the empirical background knowledge of the day already pointed towards the possibility of gill slits in the mammalian embryo. No doubt the success of the prediction counts as *some* evidence for recapitulation theory—it fits the theory so nicely. But when one properly factors in the prior probability of the predicted phenomenon, the confirmation of recapitulation is accordingly reduced.

In conclusion: this case does *not* speak significantly against the thought that 'evidence' and 'truth' are strongly correlated. It could only do so on an overly simplistic, or naïve, account of how to measure the weight of evidence for a theory. Thus I regret using this example in my *The Conversation* piece 'The Misleading Evidence that Fooled Scientists for Decades' (Vickers 2018b). As I have here argued, once one digs into the details, it is not obvious that Meckel's prediction was truly 'risky', as I stated in that article. As a piece of evidence the gill slits are only slightly misleading: although they do indeed support a radically false theoretical idea, they don't support it *very* much, given the relatively high prior probability of the predicted phenomenon in question.

In light of the analysis provided in this chapter, it is worth here reflecting on Wray's challenge to scientific realists (Wray 2018, p. 156):

> Granted, the realist may be correct to suggest that it is *mysterious* that false theories are able to make successful novel predictions. But the facts are as they are. Theories that misrepresent the world can enable and sometimes have enabled scientists to generate vindicated predictions of novel phenomena.

But the Meckel case shows that it is *not* mysterious. When judging the significance of a successful novel prediction, a great deal depends on the

[27] Brush (2015, p. 81ff.) discusses some similar cases, where a successful prediction didn't have a big impact.

wider context of empirical knowledge and scientific theorising. We can fully understand how Meckel was able to successfully predict the existence of gill slits on the basis of a radically false theory of biological development: his theory was tailored to do so, since it was tailored to accommodate a wealth of similar empirical phenomena (as detailed in Section 5). This understanding doesn't at all rule out the possibility that the only reasonable explanation of some *modern-day* prediction is that the theory in question is at least approximately true.

The analysis provided in this chapter also serves to debunk another (closely related) strategy favoured by scientific sceptics. The strategy is to first note that theories in the history of science have sometimes been 'successful', though radically false. The next step is illustrated in Stanford (2020) as follows:

> *Whatever* alternate explanation(s) the realist endorses for the success of theories that have ultimately been rejected or abandoned will also serve as a perfectly plausible alternative [to a realist explanation of the success of current theories]. (original emphasis)

However, in the Meckel case the 'alternate explanation' my analysis has uncovered—detailed in the previous section—will probably *not* 'serve as a perfectly plausible alternative' to a realist explanation of the vast majority of contemporary cases. That move assumes that one can generalise across scientific cases far more than is actually possible (Asay 2019).

In addition, it is dangerous to base the identification of future-proof science on a single piece of 'amazing' evidence, or even on a single *type* of evidence. Recapitulation theory was never a serious candidate for future-proof science, since it always suffered from numerous problems, both empirical and theoretical. There was never anything like a scientific consensus, nor would there have been even if (counterfactually) Von Baer had not existed, and Meckel's prediction of gill slits had been confirmed a decade earlier.

It is now important to ask whether one can *ever* have a 'silver bullet' piece of evidence for a scientific theory, a case where the 'no miracles argument' can be readily applied. It's important, since if silver bullet pieces of evidence don't actually exist in science, and one simply has to assess a *large body of evidence*, then it threatens to become impossible for non-scientists to digest the relevant evidence (and very difficult even for practising scientists). Novel predictive success still seems like the best candidate for such a silver bullet piece of evidence. It would have to be an *extremely* bold, risky prediction, the success of which really *would* be completely inexplicable unless the theory in question were true. Given what we've learned in this chapter, it would also have to be the case that the predicted phenomenon were not even weakly 'expected' purely on the basis of already-known phenomena. A possible case is investigated in the next chapter.

4

The *Tiktaalik* 'Missing Link' Novel Predictive Success and the Evidence for Evolution

1. Introduction

In the last chapter we saw that novel predictive success need not strongly confirm even the *working parts* of a theory, even when the prediction is prima facie 'bold' or 'risky'. So much depends on the wider empirical and theoretical context. The existence of many such cases (Vickers 2013a) presents a problem even for those who would make a defeasible commitment, asserting that novel predictive success is merely a *good guide* to truth. In addition, in this book, I'm setting a high bar: I'm not merely looking for scientific ideas that are *probably* true, but for those that are *true beyond all reasonable doubt*, and deserve to be called 'facts' (e.g. the 30 examples given in Chapter 1). If 'novel predictive success' isn't sufficient for (even approximate) truth, it's hard to see what is, without more vaguely appealing to 'sufficient success'. If we move to a vaguer construct, for example focusing on theories that are 'mature and genuinely successful' (Psillos 1999, p. 103), it becomes far less clear when the threshold for rational belief has been reached without having to simply 'ask the scientists'. And we may have a strong desire to base our beliefs directly on evidence: when it comes to persuasion, there is all the difference in the world between a 'trust me' approach and a 'look at *this* and see for yourself' approach.

What's at stake here is a crucial dilemma: to motivate belief in (and/or trust of) scientific claims by appeal to either (i) first-order scientific evidence, or (ii) grounds for trusting the evidential judgement of the scientific community. Given the quantity and complexity of the evidence in any given case, option (i) seems to require that certain individual pieces of evidence are epistemically

Identifying Future-Proof Science. Peter Vickers, Oxford University Press. © Peter Vickers 2023.
DOI: 10.1093/oso/9780192862730.003.0004

highly motivating, and novel predictive success (what else?[1]) has sometimes been put forward to play this role. In light of the historical challenges, it is now clear that novel predictive success is often insufficient to motivate belief, but a 'localist' (Asay 2019) or 'particularist' (Stanford 2021) would insist that the historical cases in the literature do nothing to show that novel predictive success could *never* be highly motivating, evidentially speaking. Instead, one might infer from the Meckel case (and others like it) merely that *that kind* of novel predictive success, in *that kind* of theoretical context, doesn't make for strong evidence. One may point to a contrast case to indicate that novel predictive success sometimes *can* be highly significant, perhaps using a famous case such as the 1919 solar eclipse evidence for general relativity. Stanford (2009, p. 384) briefly notes that it might be possible for the realist to identify 'criteria of novel prediction' that would serve to separate the predictions that are really significant from the ones that are not. Ambitiously, one could start listing the required 'criteria of novel prediction' by working one's way through the list of 20 examples in Vickers (2013a). At the end-point of this process one might suggest, 'if a theory delivers novel predictive success, and all of the conditions [. . .] are met, *then* it would be a miracle if the relevant theoretical ideas were not at least approximately true.'

However, we don't now have all of these conditions in hand, and it might be impossible to collect them all together anyway—the noted research programme might be doomed to failure. A 'hyperlocalist' (Asay 2019, pp. 602–3) or 'radical particularist' (Stanford 2021) would think so, claiming that every case is different and has its own nuances, such that only a scientist looking at the case 'from the inside' could be trusted to comment on the evidential weight of the predictive success. And in fact, *whether or not* one can collect together all such caveats, we do expect the scientific community to be a very good judge when it comes to evidence. In the search for silver bullet evidence, it thus makes good sense to look at the individual pieces of evidence picked out by the relevant scientific community when they want to persuade a reader that a given contemporary theory is true.

A good candidate theory is the theory of evolution. Assuming scientists are good at judging evidential weight, this seems like an excellent place to look for the kind of silver bullet pieces of evidence which, even by themselves, can do a lot of work to convince an open-minded audience. Certainly the scientific

[1] There have been many debates about the nature and evidential significance of successful predictions in science, which we needn't get into here. Most philosophers agree that predictions can be epistemically highly significant, either directly, or else because they are 'symptomatic' of something else that is itself highly significant (see Barnes 2008, pp. 24–7, for useful discussion).

community has sometimes found itself under great pressure to present convincing evidence for evolution. No pressure could be greater than in the context of the battle in the US over whether 'Creationism' should be taught alongside evolution theory in the schools. Looking at this case, scientists *are* attracted to novel predictive success. At least, this is what we find when we turn to the highly respected National Academies publication *Science, Evolution, and Creationism* (Institute of Medicine 2008). The prediction chosen is that of the 'missing link' between fish and amphibians, predicted to be found in rocks roughly 375 million years old. The predicted fossil was discovered in 2004, and the species was named *Tiktaalik*. The authors write, 'The discovery of Tiktaalik [was] critically important for confirming predictions of evolution theory' (Institute of Medicine 2008, p. 3). But it must be asked, how far does this piece of evidence for evolution really go? Does it have *much* persuasive power, in the grand scheme of things? We will see that things are not straightforward.

2. *Tiktaalik*: An Impressive Novel Predictive Success of Evolution Theory?

Twenty years ago there was a gap in the fossil record between fish living 380mya (million years ago), and amphibians living 365mya. At least, it's a gap if you believe in the theory of evolution. If you believe the theory it ought to be the case that organisms half-way between fish and amphibians existed between 380 and 365mya. As our understanding of the fossil record gradually improved over the course of the 20th century, the justification for believing that there were no land animals prior to (approximately) 365mya increased, to such an extent that it came to be widely accepted. Thus we are led to the hypothesis that organisms evolved to live on land during that time, between 380 and 365mya.

This led to a definite prediction: it ought to be the case that if we search thoroughly in rocks dated to 370–375mya, we will find fossils of organisms half-way between fish and amphibians, possessing all sorts of 'intermediate' features. This is prima facie a bold prediction, or, rather, a whole set of predictions. If the theory is true, and we've got the fossil record right, it is virtually guaranteed that we should find relevant fossils, in rocks of the right age, if we look for long enough. The set of predictions were not risky in the sense that predictions in physics are risky: they were not highly quantitative. However, they were still risky in the sense that one could be quite specific

Figure 4.1 *Tiktaalik*: half-fish and half-amphibian

about the age of the strata in which the fossils would be found, and a range of different anatomical characteristics that were expected. Scientists could also have confidence, given everything we know about fossil formation, that it is vanishingly unlikely that such fossils would never have formed, or would have all been destroyed. If the assumptions are correct the predicted fossils almost *have* to be there.

A project was funded, and such a fossil was indeed found, just as predicted, in 2004 (Figure 4.1). It is about as perfect an intermediate between fish and amphibians as one could hope to find. It had gills, but also primitive lungs. It had a body that is very fish-like, but unlike other fish it had a neck, and its head looked like a salamander, or a crocodile. It had special bones in its fins that would allow it to hop around on land, just as mudskippers do today. It had scales and fins, but unlike a fish it had a wrist joint, a primitive ear, and rib bones. And (of course) there are numerous other, more technical respects in which the fossil can be recognised as an intermediate between fish and amphibians (Shubin et al. 2006). Such transitions can often be represented diagrammatically. To give just one example of many (Figure 4.2), the pectoral appendage of *Tiktaalik* is 'morphologically and functionally transitional between a fin and a limb' (Shubin et al. 2006, p. 764). One doesn't need to be a scientist to see that the *Tiktaalik* appendage sits nicely between the fin and the limbs. Importantly, the *Tiktaalik* appendage in Figure 4.2 was *predicted* (even if Figure 4.2 itself was prepared after the discovery). And at least 10 other examples could be presented here: specific anatomical respects in which the 'missing link' was both *expected* to be intermediate between the fish and the amphibians, and then was *found* to be intermediate between the fish and the amphibians.[2]

[2] It would surely be ad hoc to insist that only quantitative novel predictive successes should 'count': qualitative predictions can be risky too. In addition, whilst the highly quantitative predictions come out of fundamental physics, this seems like a *poor* place to look for scientific theories that are epistemically secure. See Chapter 6 for discussion.

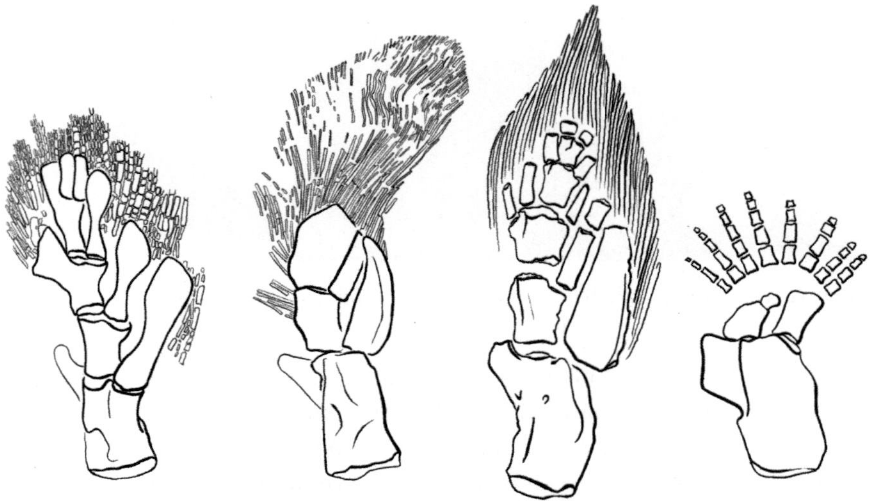

Figure 4.2 The evolution of the pectoral appendage, from fish (with fins, LHS) to amphibians (with limbs, RHS)

How should we react to the *Tiktaalik* predictive success? It initially seems like very good evidence for evolution, and it is no surprise that the National Academy of Sciences (Institute of Medicine 2008) put it front and centre in *Science, Evolution, and Creationism*. A knee-jerk reaction might bring us to declare that the relevant theoretical ideas *must* be true, that it would be vanishingly improbable ('a miracle') for those ideas to be substantially false and yet to deliver such a perfect predictive success. When the National Academies' scientists stated that 'The discovery of Tiktaalik [was] critically important for confirming predictions of evolution theory', they indicated that they were (at least back then) sympathetic to the kind of 'no miracles' thinking I am talking about—the words 'critically important' suggest that this was a *severe test* of the theory. Similarly, Ahlberg and Clack (2006, p. 748) wished to stress the importance of the result, stating that *Tiktaalik* 'demonstrates the predictive capacity of paleontology. The Nunavut field project had the express aim of finding an intermediate between *Panderichthys* [a type of fish—see Figure 4.2] and tetrapods [the earliest amphibians], by searching in sediments from the most probable environment (rivers) and time (early Late Devonian).'[3]

[3] See also Institute of Medicine (2008), p. 2: 'The team that discovered the new fossil decided to focus on far northern Canada when they noticed in a textbook that the region contained sedimentary rock deposited about 375 million years ago, just when shallow-water fishes were predicted by evolutionary science to be making the transition to land.'

But in the past decade, a great deal of caution has followed. Niedzwiedzki et al. (2010) made waves with evidence of land animals living 10–18 million years prior to *Tiktaalik*. The debate continues and the dust is far from settled, but one thing is now clear: it was never the case that a *Tiktaalik*-type fossil would *have* to be found in rocks 375 million years old. Another possibility was this: there were actually amphibians and other tetrapods living 390mya, but the relevant fossil evidence just hadn't been discovered yet. That is, there was always a possibility that the fossil record had given us a distorted view of the evolution of life out of the water. Perhaps no 'missing link' between *Panderichthys* and *Acanthostega* was needed, because *Acanthostega* actually evolved via another route, from other tetrapods already in existence 390 or 385mya. If this *did* happen, then the missing-link fossils would still be out there to be found, but not necessarily in rocks 375 million years old. The best place to look would be in older strata, in accordance with when the very earliest amphibians really *did* exist.

This reduces the significance of the *Tiktaalik* prediction somewhat. It was always consistent with evolution theory that the Nunavut prediction would be completely *un*successful. Thus the prediction—just like Meckel's gill slit prediction—actually took a probabilistic form, making it less risky and thus less evidentially significant. It was never the case that the discovery of *Tiktaalik* was 'critically important for confirming predictions of evolution theory', and this is easy to see in hindsight. It wasn't a 'severe test' in the sense of Mayo (1991).[4]

Even with the uncertainty associated with the incompleteness of the fossil record, the prediction still had a good chance of being successful. If the prevailing assumption was *correct* and life did evolve out of the water between 380 and 365mya, the prediction would be highly likely to work out. But even if that assumption was *in*correct, and life evolved out of the water several million years earlier, there remained a good chance that transitional fossils would be found in strata dated to 375mya. This is because a 'transitional' species obviously needn't go extinct once the transition has occurred: it might well remain a successful organism, suited to its ecological niche, surviving for many millions of years. Indeed, some such organisms are with us today, and at least

[4] Cf. Cleland's example of a very similar case in the context of the K-Pg mass extinction (Cleland 2013, p. 5). A 'prediction' was made in the early 1980s concerning ammonite fossils, based on the popular 'Alvarez' theory of the dinosaur extinction phenomenon. The prediction sort-of worked out, but as Cleland notes, it wouldn't have harmed the theory if it hadn't: 'This point is underscored when one considers that failure to find them [ammonite fossils] in the sea-cut cliffs of southern France wouldn't have counted much against the Alvarez hypothesis either; the local ecological crisis that decimated them in Spain could easily have extended that far north.' See Chapter 7 for detailed discussion of the Alvarez hypothesis.

one extant organism is reminiscent of *Tiktaalik* in some respects, the alligator gar, which combines functional gills and a functional lung. Thus even if the earliest transitional organisms in between fish and amphibians were alive 395mya, they might well appear in rocks dated to 375mya. At the same time, a lot can happen in 20 million years, so if the prevailing assumption was indeed *in*correct, one would need a slice of luck for the prediction to work out.

We know that the prediction was successful. But this doesn't help us decide between the two stories: (i) where the prevailing assumption was correct, and (ii) where the prevailing assumption was *in*correct but the scientists got a bit lucky. However, either way, the successful prediction surely counts as evidence for the theory of evolution. That theory (in conjunction with a few background assumptions) clearly predicts that half-fish-half-amphibian organisms like *Tiktaalik* did once exist, and relevant fossils are 'out there' waiting to be found. Even a very cautious member of the funding body considering whether to fund the 2000–4 Nunavut project would have agreed that the relevant fossils were *very* likely to exist, and the Nunavut site was a good place to look for them.

We may still ask just how strong this particular piece of evidence is for the theory of evolution. In light of this successful prediction, could anyone reasonably resist the claim that evolution is a fact? A sceptic might suggest the following:

(i) *Tiktaalik* looks like good evidence now, but so did pieces of evidence in the past that were ultimately revealed to be misleading once an unconceived alternative theory was identified.
(ii) The prediction is not as impressive as it first appears—it had a high prior probability, given the empirical knowledge already in hand.

These will be taken in turn.

First (i), could the evidence be misleading? Two very different strengths of 'misleading evidence' need to be sharply distinguished: (a) where we find that the evidence (strongly) supported a false theory, and (b) where we find that the evidence isn't as strong as it initially appeared. The cases of misleading evidence in the literature (Vickers 2018b; Frost-Arnold 2019) are usually presented as cases of the former kind, for example the Meckel case discussed in Chapter 3. It will turn out that *Tiktaalik* is indeed misleading as a piece of evidence, but only in the *latter* sense: it is less significant than it originally seemed, such that it is nothing like a silver bullet for the truth of evolution, but that doesn't mean that it isn't *evidence* for evolution, and it shouldn't for a moment make us doubt the truth of evolution.

Why isn't *Tiktaalik* as significant as it originally seemed? We've seen one reason already: there was no guarantee that it would be found at the Nunavut site, because the prediction was probabilistic. But there is another reason, and this is where option (ii) for the sceptic comes in: the *Tiktaalik* phenomenon—just like the gill slits of Chapter 3—had quite a high prior probability given the empirical knowledge of the day. And, because of this high prior probability, its power to confirm the relevant theoretical ideas is correspondingly reduced. After all, it isn't impressive to predict something that is already expected.[5]

How would such a 'high prior probability' story go? It is certainly true that scientists knew a great deal about the relevant period prior to the Nunavut project of 2000–4. They knew about a whole series of fish (*Glyptolepis*; *Gooloogongia*; *Megalichthys*; *Eusthenopteron*; *Panderichthys*) living around 385mya that were gently approaching primitive amphibians (*Acanthostega*; *Ichthyostega*) on the other side of the divide in at least some respects. Now imagine a serious and dedicated 'explorer of the phenomena', an 'empirical scientist' who proceeds without ever thinking about the theory of evolution. With intimate knowledge of the species on either side of the divide and the way they relate to each other, this empirically minded scientist might half-expect there to be other species in the fossil record, functionally and morphologically 'in between' the fish and the amphibians. Furthermore, the scientist would have a good sense of where to look, given the strata where the other species were found (without for a moment assuming macroevolution).

If our empirical scientist were to propose a very expensive project to find such a fossil, on empirical grounds and without once mentioning evolution, doubts might be raised. To justify a project such as Nunavut—lasting for several years, and involving expensive travel arrangements[6]—there would have to be extremely strong reason to expect such a fossil would indeed be found. And it might be suggested that the seven species noted above, and their morphological and anatomical relationships with each other, provide only a weak reason to expect there to be *another* species in the hypothetical 'gap' between the fish and the amphibians. Without the theory of evolution we just don't have a good enough reason to expect to find such an intermediate case, so it might be claimed.

[5] This is of course written into the Bayesian machinery: to find the degree of confirmation one divides by the prior probability. Thus the higher the prior probability, the lower the degree of confirmation for the theory. But one doesn't need to invoke Bayesian confirmation theory to make the point.

[6] Institute of Medicine (2008), p. 2: 'The team had to travel for hours in planes and helicopters to reach the site, and they could work for just a couple of months each summer before snow began to fall. In their fourth summer of fieldwork they found what they had predicted they would find.'

But our empirical scientist needn't give up yet; she could bolster her argument by drawing a comparison with other 'in between' cases, and there are many to choose from. The *Archaeopteryx* is famous for linking dinosaurs and birds. The *Ambulocetus* links land mammals with whales, and species such as *Basilosauridae* and *Dorudontinae* link the *Ambulocetus* with modern whales. For an example closer to home, 'Ardi' very nicely links apes living six million years ago with modern *Homo sapiens*. Eleven further 'missing link' examples (most of them discovered prior to 2004) are: (i) *Amphistium* (Figure 4.3), (ii) *Runcaria*, (iii) *Pakicetus*, (iv) *Attercopus*, (v) *Cladoselache*, (vi) *Gerobatrachus*, (vii) *Eocaecilia*, (viii) *Eupodophis*, (ix) *Anchiornis*, (x) *Thrinaxodon*, and (xi) *Hyracotherium*. Cases such as these—where intermediate species were expected, then discovered—already demonstrated, before we found *Tiktaalik*, that intermediate species are abundant and we should expect more. Our scientist can now appeal to the funder to have confidence in her proposed Nunavut project, arguing that the probability of finding an intermediate half-fish-half-amphibian in rocks dated to 375mya is very high. In other words—just as in the gill slit case of Chapter 3—there was arguably a high prior probability of the predicted phenomenon *purely given the empirical knowledge of the day*. How much work, then, is the theory of evolution really doing?

It could be objected that only with macroevolution assumed is the in between species *sure* to have existed, such that relevant fossils *will* be there in the rocks. But we've already seen that this doesn't work: *Tiktaalik* was not certain to be found in rocks dated to 375mya. This is because the apparent 'gap' in the fossil record between *Panderichthys* (and similar) and *Acanthostega* (and similar) might have been an illusion resulting from an incomplete fossil record. And indeed many scientists now think (based on

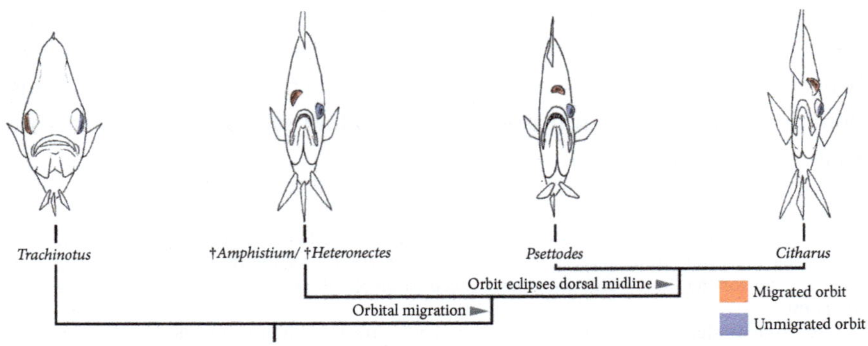

Figure 4.3 One of many 'missing link' cases

literature such as Niedzwiedzki et al. 2010) that it *was* an illusion, notwithstanding the fact that the prediction *was* successful! If we try to draw a sharp distinction between the prediction of *Tiktaalik* coming from empirical knowledge and the prediction of *Tiktaalik* coming from theory on the grounds that the prediction is more probabilistic in the former case, and less probabilistic in the latter case, we don't succeed. The difference is one of degree, and, though hard to quantify, seems to be lacking in significance.

Thus we appear to reach the following remarkable conclusion: despite appearances, the *Tiktaalik* prediction no more confirms evolution theory than the gill slit prediction confirmed Meckel's recapitulation theory. Or, more conservatively, the difference in the degree of confirmation in the two cases is not very great. It is tempting to see a dramatic asymmetry in the degree of confirmation because we know that recapitulation theory was radically false, and we know that the theory of evolution is true (when it is crudely stated, at least). But when we judge the degree of confirmation afforded to the theory by the particular prediction in question, we should work hard to avoid bias on these grounds: whether we now think the theory to be true, or false, shouldn't bear on the question of how evidentially significant the successful prediction was.

Thus we reach a conclusion that the case for evolution theory shouldn't put too much weight on the specific *Tiktaalik* prediction—it is no silver bullet.[7] And the use of this example on the first pages of *Science, Evolution, and Creationism* (Institute of Medicine 2008) should be interpreted as merely an *illustration* of the *kind* of evidence scientists are drawing on. One may wonder whether there is a better case. How about a 'missing link' novel predictive success, like *Tiktaalik*, but where the relevant empirical knowledge was much more limited, such that there wasn't a high prior probability of the phenomenon? It's not hard to find such a case if we look at the early/mid-20th century, when the empirical knowledge was much thinner. However, with limited knowledge of the fossil record—whilst this does indeed reduce the prior probability of any prediction—we can no longer make a *specific* prediction concerning the age of the rocks in which a certain kind of fossil should be found, nor the more specific features of the fossil that should be expected. And it is these specific details that made the *Tiktaalik* prediction so noteworthy.

[7] Cf. Cartwright et al. (forthcoming): '[Y]ou may be impressed by a new result—say a wonderfully precise novel prediction born out in a carefully conducted experiment—and take it to confirm your favoured theory. But whether it does what you think it does depends on a host of other assumptions being true, other experiments having been well conducted, other concepts being true to the world, and so forth.'

This isn't to say that 'missing link' predictions are not evidentially important; far from it. According to the theory of evolution, every single organism—both extinct and extant—has its predecessor 'link' ancestors. That equates to many millions of separate predictions, all of which surely go some way—once confirmed—towards supporting the truth of evolution. But in now making reference to millions of predictions, we've taken a big step towards simply pointing to the full mass of evidence. This will be a disappointing result for many, who had hoped to persuade the sceptic, or open-minded teenager, of the truth of evolution by presenting first-order evidence (such as *Tiktaalik*) to them. This was considered preferable by far, compared with (i) merely asking the sceptic/teenager to *trust the scientists* that the evidence is too strong to deny, or (ii) simply urging the sceptic/teenager to study the first-order evidence for a few years (at least) until she has an evidential overview comparable to that of at least a junior scientist. Our protagonist hoped she could take a first-order-evidence approach, since option (i) seems so unlikely to have persuasive force, and option (ii) seems hopeless because, for the vast majority, it requires too much effort. We are close to despairing that it simply isn't possible to convince the vast majority of sceptics, excepting the minority who are just *weakly* sceptical, and those who are willing and able to study the science very seriously for a number of years (some of those open-minded teenagers, perhaps).

Two remarks are in order at this point. First, it isn't my project in this book to present a strategy for persuading sceptics or teenagers. In fact, it is often the case that people are less likely to be persuaded by evidence, and more likely to be persuaded by political affiliations, rhetoric, social media, and the like (cf. Oreskes and Conway 2010). I'm not going to get into that discussion here, since it is tangential, concerning as it does the psychology of individuals and groups. My project here is to articulate how we can identify future-proof science, and to that end evidential concerns *are* of paramount importance.

Second, our protagonist needn't despair just yet. The *Tiktaalik* result isn't a silver bullet piece of evidence for evolution—not even close.[8] But that hardly means we've exhausted the ways in which we might draw on first-order evidence in order to persuade.

[8] Cf. Cartwright et al. (forthcoming): 'In trying to understand what supports a scientific hypothesis, fixing your attention on single bits... is like trying to see how the Jacana's eggs are propped up by focussing on the small rigid twigs among all the ingredients woven together to make the nest... rather than widening your vision to make out what it takes to make them supporting.'

3. The Full Body of Evidence

Suppose we accept that *Tiktaalik* is no silver bullet when it comes to warranting belief in evolution theory. For similar reasons to those already given, we might also reason that *all* 'missing link' cases will suffer in the same way, at least if we try to place great evidential weight on them: either they will have a high prior probability in light of our empirical knowledge, or else the prediction will be unimpressive because it will be vague. Furthermore, when it comes to predictions of novel phenomena coming out of evolution theory, it's hard to see anything better than a missing link prediction. Indeed, at one time it was widely assumed that evolution theory just doesn't make testable predictions, since it is purely in the business of *explaining* phenomena (Scriven 1959; cf. Godfrey-Smith 2003, p. 72). This was countered by some scholars (e.g. Williams 1973), but the examples put forward aren't as noteworthy as missing link predictions.

It may be insisted that the missing link cases aren't supposed to be impressive *as predictions*. Rather, they are supposed to provide considerable support for evolution because evolution is the only reasonable way to *explain* the relevant phenomena. Whilst I think this is a sensible move, explanations have generally been considered to be weaker, as evidence, than successful predictions: witness the criticisms of Stanford's (2006) choice of historical cases mentioned in Chapter 3, courtesy of Votsis (2007) and Saatsi (2009), for focusing on 'mere' explanatory success, and failing to focus on novel predictive success. A similar move was made against Laudan (1981): his historical examples were criticised for their lack of significant success, especially novel predictive success (e.g. Musgrave 1985, p. 211). Somehow, explanations are too easy, compared with sticking one's neck out by making a testable prediction.

This definitely isn't to say that explanations are unimportant when it comes to the evidence for evolution. What it does suggest, however, is that if we've struggled to find any silver bullet *predictive* success, we can't expect to find any silver bullet *explanatory* success, that is, one or two instances of explanation which, by themselves, strongly warrant belief in evolution. More likely, we can tell a story to the effect that the explanatory success is of great significance because of the way it unifies a huge wealth of phenomena in a highly elegant way. But this is a very long story to tell properly, since getting familiar with that huge wealth of phenomena requires years of study.

One option is to jump to a radical alternative, where one essentially forgets entirely about the first-order evidence, instead deciding what to believe by 'asking the scientists' (to speak crudely). For example, Oreskes (2019) answers the question 'Why trust science?' by pointing to a solid scientific consensus where the scientific community is diverse in a number of respects, and thus incorporates a wide range of perspectives. In the case of anthropogenic climate change, if we ask why we should be believers, we turn not to the evidence per se—features internal to scientific practice—but instead to features of the scientific community—features *external* to scientific practice.[9] After all (one may argue), the body of evidence for anthropogenic climate change is impossibly huge and complex for any individual to digest, especially non-experts. And a *little* knowledge can be a dangerous thing, as is often seen when one asks a climate change sceptic why he/she doesn't believe in anthropogenic climate change, and one receives a certain kind of biased evidential story (perhaps including evidence that the Earth used to be warmer than it is now, or evidence that we are entering a new 'mini ice age'). What's also very important to note is that the kind of evidential story that climate change *believers* often give is *similarly* a mere sketch, scratching the surface of the body of evidence, and insufficient, on its own, to justify belief. Thus Oreskes advises us to make decisions based on the character of the scientific consensus.

This is the polar opposite view and, before we analyse its merits as a way to identify future-proof science (see Chapter 5), we ought to consider alternative ways in which we might still draw on the first-order evidence. When it comes to anthropogenic climate change, Kitcher (2019)—reviewing Oreskes (2019)—prefers to 'sketch' some 'elements of the evidence' that he takes to be especially important, including the greenhouse effect. He adds to this sketch a statement to the effect that the scientists involved have worked hard, over many decades, to establish that these elements of evidence can be trusted. And he adds to this story the points Oreskes (2019) makes, about the strength of the consensus, and the fact that the community in question incorporates a diversity of perspectives and rewards individuals for identifying errors of reasoning. In other words, his is a mixed approach, where one 'sketches' some of the most salient evidence, whilst also emphasising the attitude of the scientific community, along with epistemically relevant features of that community.

Such a mixed approach might be put forward when it comes to evolution theory. Instead of trying to find one or two pieces of 'killer' evidence, one broadly sketches the fossil record, how fossils are dated, the many 'missing link' cases, the genetic evidence for evolution, how biologists have responded

[9] Cartwright et al. (forthcoming) contrast 'internalist' and 'externalist' approaches, preferring an internalist approach. Intemann (2017) is also more on the internalist side.

to the many claims of 'irreducible complexity', and so on. One can cover so *much* evidence, because one only very briefly touches on each dimension of the overall evidential story. At the end of this story one can then add a few, brief facts about the scientific community and why its judgement can be trusted.

A worry arises that it will be easy to find holes in any such sketch of the evidence, precisely because it *is* full of holes, since that's the only way to cover so much evidence in so few words. Such a story may *be* persuasive in practice, but that's not the point; indeed, often rhetoric in the media has greater power to persuade than scientific evidence, even if the arguments are full of holes. But what I'm interested in here—and what Oreskes and Kitcher both care most about—is what really *should* rationally persuade us to trust a scientific claim. And a potted sketch of the evidence doesn't seem to help with that, since so often in science the devil is in the details. We've already seen this at play in the previous chapter and in the case of *Tiktaalik*: yes, one *could* quickly skim over the *Tiktaalik* case, presenting it in 'no miracles' terms: *'In light of this perfect prediction, the theory must be true!'* And some listeners might be persuaded by that story. But this would be misleading.

An alternative option is a three-pronged attack, combining (i) a couple of *detailed* evidential stories with (ii) a broad-brush overview of the full body of evidence, and (iii) a discussion of the scientific consensus. The thought here is that one properly and fully presents a couple of pieces of evidence, to give the audience a rich sense of how evidence works in the given context, and the hope is that this then makes it rational to accept the broad-brush account of the rest of the evidence. The thought is that in seeing the full picture in a couple of key instances, one has a sense of how the gaps *could* be filled in vis-à-vis the rest of the evidence (were one to go ahead and properly study the evidence for several years). Not only this, but in the couple of detailed stories provided one sees the scientific community at work, checking and re-checking every step of the argument (cf. Chang 2004, pp. 45, 213, on 'epistemic iterations'). One can then reasonably expect that this same process of vetting took place throughout the whole body of evidence. In the case of evolution theory, this might mean providing:

(i) two detailed evidential stories, e.g. (a) the *Tiktaalik* case, and (b) scientists' understanding of why the human eye is not irreducibly complex;

(ii) a broad-brush overview of other missing link cases, other alleged cases of irreducible complexity, genetic evidence, evidence from biogeography, and so on;

(iii) evidence of a solid scientific consensus (e.g. the Pew Research Center finds that at least 97% of scientists accept evolution), together with relevant features of the scientific community (e.g. that it encompasses a large diversity of perspectives).

Indeed, the combination of (i), (ii), and (iii) just *is* what many of us have to hand if we are pushed to articulate why we believe in evolution. If people typically *do* draw on these three factors to justify belief in evolution, one would hope that they are jointly sufficient.

But one may wonder exactly what work (i) and (ii) are doing here. The examples in (i) provide scant evidence on their own, and the evidential overview provided in (ii) has several problems. For one thing, some of this evidence will be impossible to effectively summarise (e.g. genetic evidence). For another, it isn't clear that the detailed accounts (a) and (b) will go far to make it rational to accept the broad-brush evidential overview. And, even if one does accept it, how is one to judge that the weight of evidence has crossed a threshold, such that it would be completely unreasonable to doubt the truth of the theory?

It is precisely here that (iii) comes into play, of course. We judge the full weight of evidence by turning to the scientific community, and considering the strength and significance of the consensus. However, if my question is simply 'Should I believe in evolution?', it would appear that (iii) is *all* I need. The evidence covered in (i) and (ii) merely gives me a *sense* of the evidence the scientific community has built its consensus upon. But whilst this might be nice to have—it gives me a warm, fuzzy feeling inside—that evidence should not play a role in persuading me, since it remains so impoverished compared with the full evidence base as documented in the scientific journals. To give a specific example, I get a *sense* of understanding from Cleland's (excellent) summary of the reasons for believing that an asteroid impact caused the dinosaur extinction (Cleland 2013, pp. 4–5). But that summary, however well done, however apparently convincing, shouldn't persuade me. That same year (2013) a paper came out in *Geological Magazine* arguing that the Chicxulub asteroid impact occurred more than 100,000 years prior to the mass extinction (Keller et al. 2013). (See Chapter 7 for a full discussion of this case.)

Thus, were I even *nearly* persuaded by (i) and (ii), I would be too hasty, since I would be nowhere near the threshold for rational acceptance that the history of science—with all its twists and turns—has shown to be required, and that the modern-day scientific community demands.[10] And it is surely important not to convince ourselves that we justifiably believe a theory on the basis of our knowledge of the evidence, if in fact our own personal knowledge of the evidence is nowhere near sufficient to justify that belief.[11]

We reach the thought that perhaps Oreskes (2019) gets it right. A rich appreciation of the solidity of a scientific consensus together with an understanding of the make-up and dynamics of the scientific community may well be the best way to justify belief in a scientific claim. Of course, Oreskes is interested in the question 'Why trust science?', and I am interested in how we can identify future-proof science. These two projects can come apart, and, indeed, they do come apart for Oreskes who (as we have already seen) seems to be very sceptical about the possibility of future-proof science, given the lessons of history: she starkly writes that 'scientific truths are perishable' (2019, p. 50). But for me—as will soon be better explained—Oreskes' answer to the question 'Why trust science?' is an excellent taking-off point for an answer to the question of how to identify future-proof science.

In the next section we will consider how such an approach might go in the specific case of evolution.

4. The 'Consensus Approach' to Evolution

According to Oreskes (2019), the best answer to why we should trust a scientific claim focuses on aspects of the relevant scientific community, as opposed to focusing on the first-order scientific evidence. Could such an approach be effective for warranting a belief in the truth of evolution?

A sceptic casting a cursory eye over the history of science may claim the scientific community has gone wrong in its judgements time and time again. She may insist that a consensus formed in the scientific community concerning the caloric theory of heat, the phlogiston theory of combustion, and the aether theory of light, but these all turned out to be radically false theories. She may insist (quoting Kelvin, Michelson, etc.) that scientists in the 19th century thought that physics was almost 'completed', with only a few tidying-up jobs

[10] Crucially, we do want *scientists* to continue to focus on first-order scientific evidence, and to sometimes be persuaded by that evidence. See the 'Objections and Replies' section in Chapter 9 (Section 2.1) for further discussion of this important point.

[11] Cf. Allchin (2020, p. 432). See Section 3 of Chapter 9 for further discussion, including implications for school education.

left to do, but of course we all know that this turned out to be dramatically mistaken. She may also point to continental drift theory in the 20th century (see Chapter 5), and how this was 'proven' to be not only false, but impossible, and was then later accepted as not only possible, but true! She may also point to the story of the 'mesosome', an entity that, it has been claimed, 'ended up an artifact after some fifteen years as a fact' (Rasmussen 1993, p. 227) (see Chapter 8 for detailed discussion).

However, as already discussed in Chapter 2, this would be far too hasty. For one thing, non-classical theoretical physics is a special case. At one time, scientists assumed that our everyday 'classical' concepts could apply in every domain, whereas now it is appreciated that our concepts apply extremely poorly (if at all) in the quantum context (cf. Cushing 1991). This means that the problem of unconceived alternatives bites hardest in the context of non-classical theoretical physics—we can't conceive of all 'reasonable' alternatives when our best theories of theoretical/fundamental physics are so *un*reasonable from the point of view of common sense. In fact, there may be alternatives that are not only unconceived but are unconceiv*able*, since our concepts fail us quite dramatically when it comes to fundamental physics.

In addition, a distinction needs to be drawn between a case where a theory is widely considered *the front runner*, and a case where there is a very strong scientific consensus that the theory is in fact true (save for minor adjustments). When it comes to caloric and phlogiston, there were always significant problems with these theories, problems that convinced a significant proportion of the community to withhold judgement on the question of the *truth* of the theory. For example, phlogiston had its infamous 'negative weight' problem, a sufficiently thorny problem to put off a significant proportion of the community.[12] Any suggestion that phlogiston theory was considered basically true by the overwhelming majority of the scientific community cannot be supported.[13] Similarly for the mesosome case.

Contrast this with the case of interest in this chapter—evolution theory. It could never be suggested that evolution theory is merely 'the front runner'.

[12] For discussion, see Musgrave (1976); Pyle (2000); Carrier (2004); Ladyman (2011); and Best (2015).

[13] In Chapter 2, I mentioned the Newtonian, corpuscular theory of light, as another such example sometimes discussed in the literature, but where there wasn't really a 'solid' scientific consensus. At this stage in the book I have stated, more than argued, that such historical cases can be dismissed, and the work done by the word 'solid' requires greater depth of discussion. For the arguments and the details, see Chapters 5, 7, and 8.

In this case we have an extraordinarily strong scientific consensus that the theory is true (save for details). Indeed, the consensus is so strong that scientists are motivated to come together en masse and formally announce that there is a community consensus, even using the words 'true' and 'fact'. For example, in *Science, Evolution, and Creationism* (Institute of Medicine 2008) we find the following:

> [T]here is no controversy in the scientific community about whether evolution has occurred. On the contrary, the evidence supporting descent with modification, as Charles Darwin termed it, is both overwhelming and compelling. [...]. Because of this immense body of evidence, scientists treat the occurrence of evolution as one of the most securely established of scientific facts. (p. viii)

In the main body of the document we also find:

> In science, a "fact" typically refers to an observation, measurement, or other form of evidence that can be expected to occur the same way under similar circumstances. However, scientists also use the term "fact" to refer to a scientific explanation that has been tested and confirmed so many times that there is no longer a compelling reason to keep testing it or looking for additional examples. In that respect, the past and continuing occurrence of evolution is a scientific fact. Because the evidence supporting it is so strong, scientists no longer question whether biological evolution has occurred and is continuing to occur. (p. 11)

These statements (and several others I might have quoted, with a similar message) were reviewed and endorsed by the members of the Council of the National Academy of Sciences 2007–8, and the members of the Council of the Institute of Medicine 2007. This amounts to 40 esteemed scientists who are very well placed to assert that they speak for the scientific community at large. They do fall short of making the very strong statement that *no* serious scientist doubts the theory, preferring to make reference to 'the overwhelming majority of the scientific community' (p. vii; see also p. 3). Nevertheless, their statements vis-à-vis the stance of the scientific community are exceedingly strong.

What should we make of such a declaration of confidence, and the use of words such as 'true' and 'fact'? I submit that the sceptic will find *no example in the history of science* where scientists have expressed such confidence, at the community level, in a large and diverse community, and have then subsequently turned out to be dramatically mistaken.[14] Without comparable historical precedent, the sceptic's argument is considerably weakened. How might the sceptic respond?

It must be emphasised that scientists are making corrections to their claims all the time, sometimes quite significant corrections—even in *Science, Evolution, and Creationism* (Institute of Medicine 2008) *Tiktaalik* is placed squarely between *Panderichthys* and *Ichthyostega* (see p. 3) whereas (as already discussed) research since 2008 has put that idea into doubt. But this is beside the point: the document does *not* express overwhelming evidence for more specific claims about *Tiktaalik*. Adjustments in our thinking about *Tiktaalik* are fully compatible with the absolute truth of evolution. In a similar dialectic to this, Wray (2020) responds to my review of his book (Vickers 2020b), and in particular my claims about our secure knowledge of the inner and outer core of the Earth, by pointing out that 'our theory about the core of the earth is not settled', citing Irving (2018) and Grocholski (2019). But the latter articles *do not for a moment* cast doubt on 'basic' or 'crude' claims about the core of the Earth, such as that it is comprised of a solid inner core and a liquid metal outer core. Instead, they concern *details* that scientists were always open-minded about, given the state of the evidence. Precisely as we would wish. Legitimate scientific doubts about these details should have no bearing whatsoever on how we assess more crudely formulated scientific truth claims.[15]

Does the sceptic have a better response, then, to the claim that we should rationally believe a scientific theory to be future-proof whenever the relevant scientific sub-community is as certain as we find with the theory of evolution? Two options for the sceptic need to be considered:

(i) There isn't the community consensus there appears to be. The 40 scientists behind the Institute of Medicine (2008) document cannot be trusted when they make claims about 'the scientific community'.

(ii) Although there *is* a strong consensus, as claimed, this shouldn't persuade us. The scientific community is subject to biases, the bandwagon effect, and similar epistemic vices.

[14] This crucial claim will be revisited and substantiated over the coming chapters, especially Chapters 5, 7, and 8. For one thing, it's important to respond to the challenge that previous generations of scientists could have said the same thing, comparing their own (ultimately mistaken) consensuses with the consensuses of even earlier generations.

[15] Both Irving (2018) and Grocholski (2019) emphasise how difficult it is to ascertain many of the *details* concerning the properties of the inner and outer core of the Earth. Needless to say, we are very far from any scientific consensus concerning many of these details, given the current status of the evidence.

Claim (ii) will be analysed in Chapter 5, in a discussion of continental drift. For now, consider claim (i), that, when it comes to evolution, there isn't a 'solid' consensus after all.

The Institute of Medicine (2008) document falls short of stating that *every* scientist considers evolution a 'fact'. This would indeed be too bold, since these days there are literally millions of scientists around the world. Instead they choose to say that the 'overwhelming majority' of scientists believe the theory. The sceptic might try to make something of this. She might insist that this particular document (Institute of Medicine 2008) is heavily biased, and even has a political agenda, to explicitly argue against teaching Creationism in schools. With such a bias—so the objection goes—there is bound to be a temptation to exaggerate certain claims in an effort to persuade the reader. And one sure way to persuade is to use the bandwagon effect, to make out that more or less everyone else already agrees. Thus, the story goes, it is likely the authors have exaggerated when they use the term 'overwhelming majority'. More likely, our sceptic claims, there is a *mere majority* in favour of evolution theory—certainly more than 50 per cent, but not enough to call it a 'solid' consensus. The size of the dissenting sub-community, it might be claimed, is not insignificant, even if it is only 5 per cent; 5 per cent is a *lot* of scientists.

One way to test the claim that there is a consensus is to look to the journals: Are there any/many journal articles doubting evolution, or defending an alternative theory? If one looks to the professional peer-reviewed journals concerned with topics in the biomedical and biological sciences, amongst the millions of articles you will find *not one* properly peer-reviewed article doubting evolution theory.[16] Over the past several decades, just *one* paper arguing against evolution was published by a reputable science journal. This paper sneaked in through the back door, being published without proper peer review, and it later turned out that both the author and the editor of the journal were paid significant sums by the well-known anti-evolution, conservative Christian think tank the Discovery Institute. Only one month following publication, the publisher of the journal involved—*Proceedings of the Biological Society of Washington*—issued a retraction, stating that something went wrong with the review procedure, and the paper should never have been published (see Hafer 2016, Chapter 29, for discussion).

[16] It would of course be useful to fully substantiate this with a large study, as conducted by Cook et al. (2013), with follow-up paper Cook et al. (2016), for anthropogenic global warming. But even without such a study it can be substantiated, e.g. using the information in Wiles (2010) and references therein.

Ignoring this illegitimate paper, for many decades now there hasn't been a single scientific article doubting evolution theory in any one of the many hundreds of reputable scientific journals. A sceptic might say that this *doesn't* show that all scientists believe in evolution. She might insist (as the Discovery Institute has suggested) that genuine scientists who harbour doubts about evolution are forced to publish in non-scientific journals (especially religious journals promoting 'Intelligent Design Theory', such as *BIO-Complexity*). She might claim that such scientists have good scientific ideas, but that they are thoroughly suppressed by the mainstream scientific community (who are extraordinarily closed-minded). This is also how she might respond to Oreskes (2004), who notes that between 1993 and 2003 no articles in the journal *Science* doubt anthropogenic climate change (cf. Intemann 2017, pp. 194–5)

It isn't too hard to find a list of individuals with genuine science degrees, from good universities, doubting evolution theory. The Discovery Institute has worked hard over many years to produce such a list, making the list as long as they possibly can for maximum impact and influence. As of November 2020 they have reached a little over 1000 'scientists', or at least 1000 individuals who have a PhD in science or science-related disciplines (including computer science, mathematics, civil engineering, and surgery). They all signed up to the following statement:[17]

> We are skeptical of claims for the ability of random mutation and natural selection to account for the complexity of life. Careful examination of the evidence for Darwinian theory should be encouraged.

Of course, looking at the publications of these 1000 signatories, the vast majority do research on topics not at all connected to the theory of evolution (e.g. computer science). To have reached their current occupational status, they needn't have ever thought seriously about evolution, and the evidence in favour of it.

In response, the National Center for Science Education (NCSE) has produced a longer list of scientists *all called 'Steve'* (just for fun) who reject such scepticism, and embrace evolution theory.[18] More seriously, attempts have

[17] The list of individuals who have signed up can be viewed at https://dissentfromdarwin.org/ (last accessed February 2022).

[18] For the list of Steves, see https://ncse.ngo/list-steves (last accessed February 2022). (The list includes any names that are derivatives of Stephen, such as Stéphanie.) The list also has a greater number of genuine biologists than the Discovery Institute list.

been made to rigorously decide the issue of what percentage of scientists accept evolution as a 'fact'. The Pew Research Center—a (supposedly) politically neutral 'fact tank'—calculate that either 97 per cent[19] or 98 per cent[20] of scientists accept evolution theory. They describe this as 'nearly all scientists', and yet if 3 per cent of scientists really *do* reject evolution, that's a large number. There are at least seven million scientists in the world, so 3 per cent would amount to approximately 210,000 scientists. With this in mind, the Discovery Institute's list of signatories looks set to get much longer than 1000 in the coming years—10,000 signatures should be relatively easy for them. Is this significant?

If it were indeed true that 210,000 scientists have considered the evidence for and against evolution, and judged that, all things considered, serious doubts about the theory are perfectly reasonable, then that *would* look bad for anybody searching here for an example of future-proof science. It would be hard to dismiss these 210,000 scientists as a 'mere' or 'negligible' 3 per cent of all scientists. However, that has *not* happened. Investigations into the 1000 scientists on the Discovery Institute's list of sceptics reveal that many of these scientists know very little indeed about the evidence for/against evolution theory.[21] For example, some of them are mathematicians, but there is no evidence that they know *anything at all* about the fossil record, how fossils are dated, the many 'missing link' cases mentioned earlier, the genetic evidence for evolution, how biologists have responded to the many claims of 'irreducible complexity', and so on.

In addition, it must be noted that a scientist could easily sign up to the Discovery Institute's statement without actually doubting evolution theory. The first sentence of the statement is, 'We are sceptical of claims for the ability of random mutation and natural selection to account for the complexity of life.' One might fully believe in evolution, but still sign up to this statement on the grounds that there is more to the *mechanism* of evolution than mere 'random mutation and natural selection'; other factors are certainly involved, including genetic drift (cf. Abrams 2007). Turning to the second sentence of

[19] See https://www.people-press.org/2009/07/09/section-5-evolution-climate-change-and-other-issues/ (last accessed February 2022). They state: 'The survey of scientists was conducted online with a random sample of 2,533 members of the American Association for the Advancement of Science (AAAS), from May 1 to June 14, 2009. AAAS is the world's largest general scientific society, and includes members representing all scientific fields.'

[20] See https://www.pewresearch.org/fact-tank/2019/02/11/darwin-day/ (last accessed February 2022).

[21] See e.g. 'Few Biologists but Many Evangelicals Sign Anti-Evolution Petition', at https://www.nytimes.com/2006/02/21/science/sciencespecial2/21peti.html?_r=1&oref=slogin (last accessed February 2022).

the statement, we find 'Careful examination of the evidence for Darwinian theory should be encouraged', and it is obvious that a scientist might fully believe the theory whilst agreeing that such 'careful examination' is worthwhile. For example, we've already seen in this chapter that (i) the discovery of *Tiktaalik* is evidence for evolution, and also (ii) careful examination of the *Tiktaalik* case is absolutely crucial given the numerous complexities, especially the outstanding possibility that *Tiktaalik* isn't the missing link it was originally thought to be. Thus an absolute believer in the theory of evolution—unaware of any underlying agenda—could be tempted to sign the Discovery Institute's statement, and this appears to be what actually happened with at least some of the signatories (see Scott and Branch 2009).

Thus it may be true that whilst the overwhelming majority of scientists believe in the theory of evolution, a very large *number* of academics (many thousands) counting as 'scientists' in a broad sense of the term, do *not* believe in evolution. However, it is perfectly possible for an individual to be both a professional computer scientist (or engineer, or mathematician, etc.) and also an Evangelical, very strongly committed to the literal truth of Genesis, and passionately (intuitively) against the idea of evolution. The layperson with a passion for learning about the evidence for/against evolution, and who dedicates a year (say) to reading several of the most important books and articles on the topic, will count as much more of an 'expert' than a scientist with a PhD in computer science who has *not* read any relevant literature. Thus the non-scientist's judgement of the warrant for believing evolution theory should be carefully based on a good understanding of the distribution of epistemic attitudes within the (relevant) scientific communities, including a sense of the varying motivations for those attitudes.

To sum up, there really *is* an exceedingly strong scientific consensus vis-à-vis the truth of evolution. The link between such a consensus and truth will be carefully analysed in Chapter 5.

5. Conclusion

One cannot (in practice[22]) justify a belief in evolution directly on the basis of first-order evidence, not by (i) providing a detailed discussion of one or two 'killer' pieces of evidence, nor by (ii) sketching the full body of evidence, nor by

[22] Of course, in principle one could simply study the evidence all day every day for the next 20 years. Such a remote possibility is not taken seriously here.

(iii) a combination of the two. One's justification ought to come from the full details of the full body of evidence.[23] This can only be accessed indirectly, via a study of the scientific community. Of course, to identify future-proof science one will hardly ever need to draw on the *entire* body of evidence—one will expect that, if the scientific ideas are indeed knowably future-proof, the evidence threshold was probably crossed at some point in the past, when we had *less* evidence (perhaps a lot less) than we now have. Nevertheless, if one accesses the evidence indirectly, via features of the scientific community including a solid scientific consensus, one might claim to have tapped into the full details of the entire evidence base, since the consensus has formed in light of those details. Although one may not *need* to draw on the entire evidence base, it doesn't harm to do so.

The most pressing question now concerns when a scientific consensus can be trusted. I will investigate whether it is possible to judge the reliability of a scientific consensus by considering the track-record of such consensuses over the history of science. I suggest that no consensus comparable with the scientific consensus vis-à-vis the truth of evolution has ever been overturned. This claim needs substantiating. Moreover, even once it is substantiated, it is necessary to defend against the threshold problem. This is the worry that even if today's consensuses are somehow 'better' than at any time in the history of science, past consensuses were also better than even earlier consensuses, and future consensuses will be even better than ours. How can we be confident that the character of the consensus we *now* have means that we have certainly crossed some threshold, such that the scientific ideas are future-proof? How can we be sure that consensuses are reliable, and are not simply the result of the bandwagon effect and similar effects hazardous to the pursuit of truth?

These issues will be tackled next. We begin with a study of an (alleged) scientific-consensus-gone-wrong: the case of continental drift.

[23] Unless one is a scientist; scientists should continue to focus on first-order evidence, and sometimes argue for/against scientific claims *drawing on the first-order evidence they know*. See Section 2.1 of Chapter 9 for discussion.

5

The Judgement of the Scientific Community

Lessons from Continental Drift

1. Introduction

In the last chapter I turned away from first-order evidence, and towards identifying future-proof science via features of the scientific community, especially a solid scientific consensus (a type of 'higher-order evidence'; see e.g. Feldman 2009 and Parker 2020).[1] When it comes to evolution we might usefully contrast two extreme cases:

(i) Somebody who studies the scientific evidence assiduously, learning about missing link cases, cases of vestigiality, evidence from biogeography, and so on, and convinces herself that the evidence has crossed a threshold such that evolution ought to be called a 'fact', not a 'theory'.
   - N.B. This hypothetical individual *only* knows about the first-order evidence, and has no real sense of what the scientific community thinks about the weight of evidence.[2]

(ii) Somebody who studies the scientific *community* assiduously, and concludes that evolution is a 'fact' (not a 'theory') on the basis that at least 97 per cent of scientists (and an even greater percentage of *relevant* scientists) think evolution is a 'fact'.
   - N.B. This hypothetical individual knows *nothing* about the relevant first-order evidence.

[1] Feldman (2009, p. 304): 'Higher-order evidence: Evidence about the existence, merits, or significance of a body of [first-order] evidence.' For the opposite view, arguing against a consensus-based approach, see Intemann (2017); my hope is that this book successfully counters Intemann's arguments.

[2] This is very close to what Barnes (2008, chapter 2) calls an 'individualist', except that Barnes' individualist may be aware of the judgements of others (whilst not being influenced by those judgements). His contrast is not mine, as can be seen by considering his definition of the alternative, the 'epistemic pluralist', somebody 'who counts the judgements of other agents about the probability of T as epistemically relevant to her own assessment of T' (p. 27).

Identifying Future-Proof Science. Peter Vickers, Oxford University Press. © Peter Vickers 2023.
DOI: 10.1093/oso/9780192862730.003.0005

Both of these extremes are problematic. The former case was tackled in Chapter 4: it would take a prohibitively long time to competently assess a sufficient amount of the first-order evidence, and even at the end of this long process one would be using one's intuitions to judge whether a weight-of-evidence threshold has been crossed, such that the theory is surely future-proof. The latter case is also problematic, as it stands. On what grounds could one link the confidence of the scientific community with the sure truth of the theory? Most obviously, a critic would insist that any such link would be to ignore history (Oreskes 2019, p. 50, fn. 88). Our protagonist could respond that in those cases from history the scientific community weren't *this* confident—a past scientific community has *never* been *this* confident and then ended up wrong. But this does little to help, since the sceptic will note that past generations could have said *exactly the same thing* about even earlier generations, and then subsequently their cherished theories ended up being wrong (Wray 2018, p. 94).

If we are going to opt for some variation on option (ii), it must be asked: How can we possibly know that we've crossed a threshold such that we can be sure the relevant scientific ideas are future-proof? How can we have confidence that, 1000 years from now, there won't be philosophers of science looking back, observing the collapse of evolution theory along with its replacement by a (currently unconceived) alternative, and saying something like the following:

- The scientific consensus in the year 2020 was only 97 per cent—this was one 'red flag', since the dissenting 3 per cent comprised many thousands of scientists;
- The scientific community in the year 2020 lacked diversity, thus encompassing a limited range of perspectives, and this was even noted at the time—this was another 'red flag'.[3]

So it is possible to imagine a philosopher who believes in 'future-proof science' merely raising the bar each time she comes across a counterexample. The sceptic suggests that in the future she will probably want to raise the consensus requirement to 98 per cent, for example. As the bar keeps getting raised, nobody has confidence in any given threshold that has been reached—everyone expects that the bar will probably be raised again, or at least one

[3] There is a great deal of literature on the importance of improving the diversity of the scientific community. See e.g. Moss-Racusin et al. (2016), Kang and Kaplan (2019), and the references therein.

feels unsure. This is described as 'the threshold problem' by Stanford (2009, p. 384, fn. 3).

I take the term 'red flag' from Oreskes (2019, p. 75). The idea behind a 'red flag' is that one can easily see in hindsight—and might even have seen at the time—that the science in question wasn't yet settled. One of Oreskes' key examples is Alfred Wegener's continental drift theory. This is especially interesting because of the heavy community resistance to Wegener's claim, for such a long time, before it was finally accepted in the mid to late 1960s. Frost-Arnold (2019, p. 913) invokes continental drift as one of his examples of 'misleading evidence' in the history of science, where, despite their best efforts, scientists were misled by the evidence—many scientists were sure about something that was later proven false. In this particular case, they were sure that the continents could *not* 'drift' in the way put forward by Wegener (1915). Oreskes wishes to look back at this episode and identify a red flag, such that it ceases to be a threat to our trust in contemporary scientific ideas. No doubt, however, sceptics such as Frost-Arnold would say that identifying such a red flag is easy *in hindsight*—we can always look back, pick out this or that feature concerning what was happening, and say 'this was a sign that something was wrong'. But being able to do that in hindsight provides no reason to suppose that it was realistically possible, in practice, at the time. And when it comes to our trust in *contemporary* scientific ideas, we obviously do not have the benefit of hindsight yet!

In the search for future-proof science, there is certainly much to be learned from the continental drift case. It is an education for all of us, concerning the extent to which scientific communities—even communities outside fundamental physics, where a lot of the examples usually come from—can be sure about things that are later overturned. We start in Section 2, analysing the extent to which there was a consensus against continental drift, and in favour of continental permanency (fixism). Sections 3–4 deal with the threshold problem; Section 5 concludes.

2. Was There a Consensus Regarding the Truth of Continental Permanency?

It has sometimes been said that there was a scientific consensus against continental drift for a significant period, some time between 1915 when it was first seriously debated and 1965 when drift was starting to become popular. For example, Hellman's (1998) chapter on continental drift is entitled

'Wegener versus Everybody', suggesting a very strong consensus against Wegener's ideas. And in the relevant years fixists themselves (those believing in continental permanency) sometimes appealed to a strong scientific consensus against drift, and in favour of permanency, at least as concerns *relevant* scientists. To give one clear example, the hugely influential American palaeontologist George Gaylord Simpson (1943) wrote:

> [T]he verdict of paleontologists is practically unanimous: almost all agree in opposing his [van der Gracht's] views, which were essentially those of Wegener. [...]
>
> The fact that almost all paleontologists say that paleontological data oppose the various theories of continental drift should, perhaps, obviate further discussion of this point [...]
>
> It must be almost unique in scientific history for a group of students admittedly without special competence in a given field thus to reject the all but unanimous verdict of those who do have such competence. (p. 2)

For Simpson—to put it bluntly—the only scientists favouring continental drift were those who weren't competent to comment on it, at least not with any genuine scientific authority (e.g. paleobotanists—'students' when it comes to paleontology), and whose opinions could thus be ignored.[4]

However, any claim that there was a solid scientific consensus against drift can quite easily be challenged. First, one can question Simpson's move to ignore scientists who were not paleontologists. Wegener had already written in 1929, in the very first sentence of the Foreword to the fourth, revised edition of his 1915 book,

> Scientists still do not appear to understand sufficiently that all earth sciences must contribute evidence towards unveiling the state of our planet in earlier times, and that the truth of the matter can only be reached by combining all this evidence. (Wegener 1929, Foreword)

And, indeed, one might think that Simpson could have learned this very lesson from an earlier controversy in the history of science, namely, the example of evolutionary biologists fighting against Lord Kelvin's insistence that the Earth

[4] Stewart (1986, p. 266) states that 'drift theory was... almost completely ignored and even suppressed.' I originally used the word 'suppressed' myself, but I moved to 'inhibited' on the advice of Naomi Oreskes, since 'suppressed' often means 'silenced', which is too strong.

couldn't possibly be as old as they were saying it was (Hallam 1983, p. 107ff.). In his 1943 paper, Simpson repeated Kelvin's mistake almost exactly, assuming that because *his* field strongly favoured a given theory, that ought to be the end of the matter.[5]

Second, Simpson certainly exaggerated when he referred to 'almost all paleontologists', since that isn't true when one considers scientists operating across the whole world. Most references in the literature to a 'consensus' regarding continental permanency, and against drift, are restricted to a particular country or region. For example:

> The response to Wegener in Europe was hardly more sympathetic. After some initial expressions of sympathy and interest the consensus turned against him. (Hallam 1983, p. 161)
>
> Most North American and Australian geologists prior to the middle 1960s were strongly opposed to mobilism. (Frankel 2012, p. 31)
>
> . . . the general consensus of the North American Earth science community in favor of permanentism [in the late 1940s]. (Frankel 2012, p. 116)
>
> [F]rom the 1920s through the mid-1960s most leading Australian Earth scientists were fixists. Most never argued against mobilism in print; they seemed confident that fixism was true, and saw no need to defend it or often even to mention it. (Frankel 2012, p. 496)

Taken together, one *might* reasonably refer to a scientific consensus in favour of fixism, and against drift, in North America, Europe, and Australia. Caution is required since, as noted elsewhere in this book, one should be highly cautious about a possible slide from 'most scientists' to 'nearly all scientists'. But even putting that to one side, the above quotes only support talk of a consensus against drift *in the noted regions*. Whilst this is highly significant, the key point here is that we ought to be judging the opinions of Earth scientists *right across* the world.

One thing that helps to distort our perception of the continental drift controversy is the tendency for literature to present particularly dramatic and shocking quotations, especially those from influential scientists. The quotations from Simpson, above, are one such example, combining (i) a highly respected scientist, with (ii) an extremely strong opposition to continental

[5] Wegener himself had, to his regret, previously 'in a weak moment' (Wegener 1929, Foreword) said that geo*physics* took very significant precedent in the debate about continental drift.

drift. Another familiar example from the literature concerns the highly influential and internationally respected senior British scientist Harold Jeffreys; the following provides an excellent illustration of Jeffreys' attitude to drift:

> When asked by David Brown in 1959 if he had read a paper by S. Warren Carey, an avid mobilist at the University of Tasmania who had showed rigorously that the fit of Africa and South America was excellent, Jeffreys remarked that he had no intention of reading Carey's paper. Similarly, when invited by Jim Everett to examine his impressive computerized fit of the continents, he said that he did not have enough time.
>
> (Frankel 2012, p. 23)[6]

And in the highly respected *American Journal of Science*, the American geologist Bailey Willis described continental drift as a 'fairytale', and despaired that 'further discussion of it merely incumbers the literature and befogs the minds of fellow students' (Willis 1944, p. 509). Clearly for Willis the debate should have been closed off, with no articles supporting continental drift allowed into the literature (much as we find with anti-evolution articles today).

But although these were influential figures, and these examples are illustrative of the drama of the time, these are still *only three scientists*. When we ask after a possible scientific consensus, it is a mistake to weight the opinion of highly ranked scientists above others.[7] Many historians and philosophers of science have commented on the fact that individual scientists often combine moments of 'genius' with other moments of poor reasoning, or simply poor scientific judgement, including such figures as Newton, Darwin, and Einstein. For example, Jaki (1969), discussing Olbers' paradox, writes of 'the simultaneous existence of excellence and shortsightedness in an outstanding scientist about a topic in his own field' (p. 142). In addition, it must be noted that often the most senior and respected scientists are those who are in the final years of their scientific careers, are conservative in their views, and feel particularly confident (given their seniority) about expressing their views forcefully, even

[6] Cf. WB Hamilton looking back in 2003: 'drift was rejected without consideration of evidence' (see Rossetter 2022, p. 11). Jeffreys would certainly have counted as one of Hamilton's 'uninformed gurus' (ibid.).

[7] Cf. Longino (1990), whose concept of 'tranformative criticism' within a scientific community aiming at 'objectivity' requires that 'intellectual authority must be shared equally among qualified practitioners' (p. 76).

at times patronisingly (cf. Hallam 1983, p. 162).[8] What I'm interested in here is *not* whether the 'big names of the day' rejected drift, but rather the question whether there was a solid, international scientific consensus against drift. If we're in the business of identifying future-proof science via identifying a scientific consensus, the pressing question is, would we have held up continental permanency as an example of future-proof science given the extent of the consensus?

Once one gets past the most dramatic quotations, and starts to get a sense of the whole international community, it is soon very clear that it is called the continental drift *controversy* (Frankel 2012) for a reason. The reason is simply that there *was* a debate, and there *was* disagreement in those years. This was decidedly *un*settled science, not the kind of settled state of affairs required—so this book argues—to confidently and justifiably describe a scientific idea as 'future-proof'. This can be quickly demonstrated. The following is a list of 35 examples of serious bona fide scientists *explicitly* supporting continental drift in serious scientific publications during the relevant years.[9] I have limited the list to one entry per author, otherwise it could easily be extended to 100 publications, since some of the listed authors published more than 10 articles on the topic (e.g. the British geologist Arthur Holmes):

1. 1910. FB Taylor, article in *Geological Society of America Bulletin*.[10]
2. 1915. A Wegener, book published with Friedrich Vieweg & Sohn.[11]
3. 1924. L Joleaud, article in *Journal de la Société des américanistes*.[12]
4. 1924. E Argand, article in *Proceedings of XIIIth International Geological Congress*.[13]

[8] Sometimes it has been claimed that the opinions of senior scientists are (much) more important than the opinions of junior scientists. Regarding the probability of alien life, Luis Alvarez reportedly stated in 1967: 'There is no democracy in physics. We can't say that some second-rate guy has as much right to opinion as Fermi' (Greenberg 1999, p. 43). But the continental drift controversy stands as a reductio of this kind of thinking. It has been thoroughly documented that the majority of 'big names' in the scientific community rejected continental drift even when the evidence was very strong. Stewart (1986) presents a quantitative analysis supporting this claim; he argues that, 'Scientists with more publications are less likely to accept a revolutionary theory that would undermine the intellectual foundations of their previous contributions to geology' (p. 271) and 'the more prominent geoscientists tended to oppose drift theory' (p. 274).

[9] Several significant drifters are not mentioned, including van der Gracht and Runcorn (see Frankel 2012). On Runcorn, see Hallam (1983, p. 164): 'Although Runcorn had been trained to believe, like all other British geophysicists, that continental drift was impossible, he quickly became an enthusiastic convert . . .'.

[10] See Taylor (1910). Taylor was an American geographer and geologist.

[11] See Wegener (1915). Wegener was a German meteorologist.

[12] See Joleaud (1924). Joleaud was a French paleozoographer.

[13] See Argand (1924). Argand was a Swiss geologist.

5. 1926. RA Daly, book published with Charles Scribner's Sons.[14]
6. 1926. LJ Krige, article in *South African Journal of Science.*[15]
7. 1928. EB Bailey, article in *Nature.*[16]
8. 1928. A Holmes, article in *Nature.*[17]
9. 1928. A Du Toit, article in *Proceedings of the Geological Society of South Africa.*[18]
10. 1931. A Windhausen, book published with Buenos Aires.[19]
11. 1934. R Maack, article in the *Zeitschrift der Gesellschaft für Erdkunde zu Berlin.*[20]
12. 1934. A Seward and V Conway, article in *Annals of Botany.*[21]
13. 1936. B Sahni, article in *Journal of the Indian Botanical Society.*[22]
14. 1937. SP Agharkar, article in *Proceedings of the 24th Indian Science Congress.*[23]
15. 1937. WJ Jongmans, article in *Proceedings of 2nd Congress of Carboniferous Stratigraphy.*[24]
16. 1937/8/9. FE Suess, article in *Fortschritte der Geologie und Paläontologie.*[25]
17. 1939. F Dixey, article in *Transactions and Proceedings of the Geological Society of S. Africa.*[26]
18. 1940. AI de Oliveira and OH Leonardos, book published with Rio de Janeiro.[27]
19. 1941. A Wade, article in *Proceedings of the Royal Society of Queensland.*[28]
20. 1944. DH Campbell, article in *Science.*[29]
21. 1946. J Hutchinson, book published with P. R. Hawthorn.[30]

[14] See Daly (1926). Daly was an American geologist.
[15] See Krige (1926). Krige was a South African geologist.
[16] See Bailey (1928). Bailey was a British geologist.
[17] See Holmes (1928). Holmes was a British geologist.
[18] See Du Toit (1928). Du Toit was a South African geologist.
[19] See Windhausen (1931). Windhausen was a German oil geologist.
[20] See Maack (1934). Maack was a German geologist.
[21] See Seward and Conway (1934). Seward was a British paleobotanist.
[22] See Sahni (1936). Sahni was an Indian paleobotanist.
[23] See Agharkar (1937). Agharkar was an Indian paleobotanist.
[24] See Jongmans (1937). Jongmans was a Dutch paleobotanist.
[25] See Suess (1937/8/9). Suess was an Austrian geologist.
[26] See Dixey (1939). Dixey was a South African geologist.
[27] See Oliveira and Leonardos (1940). Oliveira and Leonardos were Brazilian geologists.
[28] See Wade (1941). Wade was a British petroleum geologist.
[29] See Campbell (1944). Campbell was an American paleobotanist.
[30] See Hutchinson (1946). Hutchinson was a British botanist. In his 1946 book he was more explicit than most: 'I am a firm believer in Wegener's ideas' (p. 13).

22. 1947. R Good, book published with Longmans, Green & Co.[31]
23. 1950. EV Wulff, book published with Chronica Botanica Co.[32]
24. 1951. LL Fermor, article in *Transactions of the Institution of Mining and Metallurgy*.[33]
25. 1952. RJ Adie, article in *Geological Magazine*.[34]
26. 1953. L King, article in *American Association of Petroleum Geologists Bulletin*.[35]
27. 1953. J Walton, book published with Adam & Charles Black.[36]
28. 1955. SW Carey, article in *Papers and Proceedings of the Royal Society of Tasmania*.[37]
29. 1956. E Irving, article in *Geofisica Pura e Applicata*.[38]
30. 1959. MS Krishnan, article in *Proceedings of the International Oceanographic Congress*.[39]
31. 1961. H Martin, article in *Transactions and Proceedings of Geological Society of S. Africa*.[40]
32. 1961. E Plumstead, article in *South African Journal of Science*.[41]
33. 1961. W Hamilton, article in *Geological Society of America Bulletin*.[42]
34. 1961. JJ Bigarella and R Salamuni, article in *Geological Society of America Bulletin*.[43]
35. 1962. GA Doumani and WE Long, article in *Scientific American*.[44]

This list of 35 explicitly pro-drift publications, written by bona fide, relevant scientists, fully spans the period in question, from 1910 to the early 1960s. What is more, for every one scientist willing to *explicitly* support continental drift in print, there were a great many more who were curious about

[31] See Good (1947). Good was a British botanist.
[32] See Wulff (1950). Wulff was a Russian palaeobotanist.
[33] See Fermor (1951). Fermor was a British geologist.
[34] See Adie (1952). Adie was a South African geologist.
[35] See King (1953). King was a South African geomorphologist. In his 1953 article he declared, 'These correspondences [between Africa and S. America] . . . can not be accommodated in any other way than by drift' (p. 2173).
[36] See Walton (1953). Walton was a British botanist.
[37] See Carey (1955). Carey was an Australian geologist.
[38] See Irving (1956). Irving was a British palaeomagnetist.
[39] See Krishnan (1959). Krishnan was an Indian geologist.
[40] See Martin (1961). Martin was a South African stratigrapher.
[41] See Plumstead (1961). Plumstead was a South African stratigrapher and paleobotanist.
[42] See Hamilton (1961). Hamilton was an American geologist. Rossetter (2022) discusses Hamilton's work and provides an extensive bibliography.
[43] See Bigarella and Salamuni (1961). The authors were both Brazilian geologists.
[44] See Doumani and Long (1962). Long was an American geologist. Doumani was a Lebanese Palestinian geologist.

continental drift, behind the scenes.[45] And for every scientist who was sympathetic, there were a great many more who were largely *un*sympathetic and yet not sure *enough* to consider continental permanency a certain fact—that is to say, they considered the debate to be definitely open, even if continental permanency was, for them, the clear front-runner. As Hallam (1983) puts it, 'A large number of people were either noncommittal or had a sneaking sympathy with the ideas of Wegener and du Toit, but considered it professionally wise to keep fairly quiet about it' (p. 174). Numerous specific names could be mentioned here, including the following:

- CR Longwell, from the USA, 'was at least sympathetic, but unconvinced by the evidence' (Hallam 1983, p. 150);
- DN Wadia, from India, 'had no preference' (Frankel 2012, p. 105);
- Hoeg, from Norway, wrote in 1937: 'we have no better solution [than drift] at present' (Frankel 2012, p. 106);
- HN Andrews, from the USA, found drift 'the most appealing explanation' of various phenomena (Frankel 2012, p. 106).

These individuals, and many others, were simply open-minded, even if they did have a (slight/strong) preference.[46]

What's particularly interesting about this case is the obvious importance of an *international* consensus vis-à-vis future-proof science. For a good period between 1915 and 1965 there really was *quite* a solid consensus in the USA, in Europe, and in Australia. This consensus was strongest for palaeontology and geology, but also quite strong for other Earth sciences. Looking at these regions alone, one could be misled into thinking that virtually all scientists agreed that continental drift was impossible and continental permanency was a fact. However, once one asks after an international consensus, and considers all different Earth sciences (especially paleobotany), such a thought immediately collapses.

[45] A crucial further point is that many pro-drift papers, submitted to journals, were rejected by anti-drift reviewers. Cf. WB Hamilton looking back in 2002: '[p]eer review then, as now, was often used to block dissemination of concepts contrary to beliefs held by reviewers, and it was often difficult in those stabilist years to get pro-drift materials published outside the rare mobilist symposia' (Hamilton 2002, p. 362; see Rossetter 2022).

[46] The American geologist Arthur Mirsky is an interesting case of a 'fixist' who later claimed not to have had a *very* strong view; he writes that at one particular conference in 1963, 'I got the urge to sound more hardline anti-drift than my actual "maybe it is so, but I am not yet convinced" tone. [...] I needled them with my apparent hard-line opposition rather than my actually more moderate, questioning view' (Frankel 2012, p. 372). If this is an honest recollection, Mirsky—although known as a fixist—would have been strongly opposed to closing off the debate, in which case he couldn't count as part of a scientific consensus asserting the truth of continental permanency.

To sum up this section, the case of continental drift was supposed to be a worry because it seemed to show that scientists can go wrong in their evidential judgements, *even* when we consider an entire scientific community. It thus threatened to undermine the strategy of identifying future-proof science by identifying a scientific consensus. However, there is a world of difference between the proffered 30 examples of future-proof science listed in Chapter 1, and the case of continental drift. The scientific debate vis-à-vis those 30 examples *has* been closed off, because the international scientific community has reached a solid consensus on those examples (though, as repeatedly stressed, there are many open questions regarding the *details* of those 30 examples). It would be absolutely impossible to provide a list of 35 serious scientific publications explicitly in opposition to any one of those 30 examples, in the way I have here provided a list of serious scientific publications opposed to continental permanency during the relevant years. Indeed, one couldn't hope to find even five such publications for any one of those 30 examples, not in serious, unbiased scientific journals, in (relatively) recent years. And that's a really big deal given that the number of scientific journal articles has a doubling rate of approximately 20 years.[47]

An obvious objection presents itself at this point. I have found a way to make irrelevant past historical cases. I've done it (taking inspiration from Oreskes 2019) by bringing in some criteria for future-proof science that were lacking in past cases: for example, there has to be a solid international consensus, with a good deal of diversity of perspectives within that international community (cf. Tucker 2003, p. 512). For example, it can't be the case that one can identify as many as 35 serious scientific publications (from 35 different authors) opposing the majority position. Now, I take it that these criteria are obviously relevant to epistemic security—they *are* reasons to increase one's confidence. But how can we be sure that they should increase our confidence *so much* that we are justified in asserting that the scientific ideas under scrutiny will still be with us in 1000 years? This is the threshold problem.

The threshold problem states that, whatever criteria we've introduced, we can't know that they are sufficient for a scientific idea to be future-proof. The two most obvious ways to respond are:

[47] One caveat here: scientific journal articles should take precedence over books. It is relatively easy to publish a book making any claim whatsoever if one is willing to pay and/or one doesn't care who publishes it. One can even self-publish, of course. The books I have chosen in my list of 35 examples of pro-drift publications are serious scientific books, published by reputable publishing companies.

(i) Provide an explanation for why the criteria are sufficient.
(ii) Decline to *explain*, and provide evidence that the criteria *just are* sufficient.

The first option is tackled in Section 3, leaving option (ii) for Section 4.

3. Tackling the Threshold Problem (i): Analysing Community Dynamics

How might it be argued that the criteria are sufficient? First let's clarify what the claim is. The claim is that we can identify future-proof science by identifying a solid international scientific consensus regarding a scientific claim. More carefully, I put forward for consideration the following two criteria:

(1) At least 95 per cent of relevant scientists are willing to state the claim unambiguously and without caveats or hedging. If prompted, they would be willing to call it an 'established scientific fact'.[48]
(2) The relevant scientific community must incorporate a substantial diversity of perspectives.[49]

I originally considered including another criterion, along the lines of 'The claim has been appropriately scrutinised', or 'The claim has stood the test of time' (assuming that more time equals more scrutiny). In fact, in Vickers (2020a) I debated how long to wait, writing 'perhaps a cautious realist should not make a realist commitment to any theory, however successful, until a few

[48] When it comes to 'relevant scientists' I let common sense dictate. Somebody publishing in mathematics journals, or engineering journals, or foundations of physics journals, wouldn't be relevant for a scientific claim concerning evolution by natural selection, for example. There will be a grey area, of course, but there are grey areas in anything and everything. When it comes to the 'relevant field of study' there will be reasonable and unreasonable places to draw the line. It could be suggested that instead of separating scientists into relevant/irrelevant, we would do better to separate them into three categories: highly relevant, relevant, and irrelevant, where the opinions of 'highly relevant' scientists are double-weighted. However, sometimes the 'outsider' scientist (who is still *quite* relevant) sees the issues in a field in an importantly unique way (e.g. Wegener), and we must be cautious about downgrading the significance of their opinions.

[49] Common sense must dictate here, and we mustn't expect precise criteria for 'sufficient diversity'. The relevant community must be reasonably balanced when it comes to dimensions such as academic specialism, gender, nationality, religion, political inclination, and so on (where some dimensions are more important than others). Communities in the 20th century have often been sufficiently diverse, as shown by the fact that 'scientific facts' have been put forward, and then subsequent work has essentially proven that those facts are future-proof. See Chapter 9, 'Objections and Replies', Section 2.2, for further discussion.

years have passed and the dust has settled' (p. 989). However, the amount of time cannot be fixed, even vaguely at 'a few years', because it depends on at least (i) the amount of effort the community is devoting to scrutinising the claim, and (ii) the kind of access the community has to the target of the claim (e.g. can they see it with an electron microscope, or is their access much more indirect?). Depending on these factors, the 'test of time' can sometimes be a matter of weeks, not years (see Chapter 8). If we include an explicit condition, insisting that an *appropriate* amount of time has passed, allowing for *sufficient* scrutiny, the match between the conditions and future-proof science becomes trivial; one could no longer use the conditions to identify future-proof science (see 'Objections and Replies' in Chapter 9).

I believe a better approach is to simply stick with two criteria. Although it's true that an appropriate amount of scrutiny needs to take place, this is something the scientific community itself judges in the process of reaching its 95 per cent consensus. So if the two criteria are satisfied, then sufficient scrutiny has taken place, so I claim. A reason to believe this is that the criteria are met for the 30 examples given in Chapter 1, and they are *not* all met for the case of continental permanence (given the prevailing views in South America, South Africa, and India). In fact, there has never been a case in the entire history of science where these two criteria *were* met, and the claim in question turned out *not* to be future-proof. In any case where the two criteria have been met—and that's a lot of cases—the claim in question has never been overturned.

Thus the criteria draw a line between the 30 examples in Chapter 1, and the case of continental permanence that was supposed to make us worry about trusting the judgement of the scientific community. However, it could be claimed that I've somewhat cherry-picked the criteria to *ensure* a line is drawn between continental permanency and those 30 proffered examples of future-proof science. A sceptic will say that I've (ad hocly) raised the threshold to just the level where one-time historical challenges are dismissed as irrelevant, but without providing any real argument that these criteria are actually sufficient for a scientific claim to be future-proof. Why, our sceptic asks, don't we need to raise the bar even higher? Why not make it 98 per cent in criterion 1? And when it comes to criterion 2, why not insist that the gender balance in the community needs to be much closer to the ideal than the status quo? Indeed, our sceptic expects that the bar might well be raised in this way at some point in the (perhaps distant) future, when future scientific revolutions have demonstrated that the two criteria are *not* sufficient for a scientific claim to be future-proof. She isn't certain about that, but she doesn't need to be certain—she just needs to cast doubt on the certainty of her opponent, who firmly believes in future-proof science. She doesn't *deny* that many current scientific claims are future-proof; she simply claims that we can't know. Thus she thinks her position is modest, and her opponent's position is hubristic.

How can our believer be so sure? One option is for her to appeal to examples from the history of science which apparently show how incredibly hard-won a solid international scientific consensus is. One such example is the debate over atomism (Gardner 1979; Chalmers 2009; Ivanova 2013). The idea is that one can see, in examples such as this, that any solid international scientific consensus is *so* hard-won that the evidence base has to be truly enormous to achieve it. And, with an evidence base *that* enormous—so the argument goes—the threshold for truth simply must have been comfortably crossed. Considering the atomic hypothesis, if we start the story in the mid-19th century, there were clearly widely different attitudes in the relevant scientific community. Some scientists were quick to embrace atomism on rather limited evidence, whilst others were extremely reluctant, and the majority fell somewhere in between. As the evidence accumulated, some scientists got on board quickly, but others put up quite a fight. Our protagonist claims that the conservatives were only won over when the evidence had clearly crossed the threshold for truth—when every possible loophole and argument had been thoroughly explored.

Studies on the uptake of new ideas generally have been ongoing for decades, and show that one can model the speed of uptake across the relevant community quite accurately with a simple bell-curve/normal distribution (Figure 5.1). This can be applied in a very wide range of contexts, whenever and wherever a community is faced with the question of how quickly (if at all) to leave the status quo behind and get on board with new ideas, technologies,

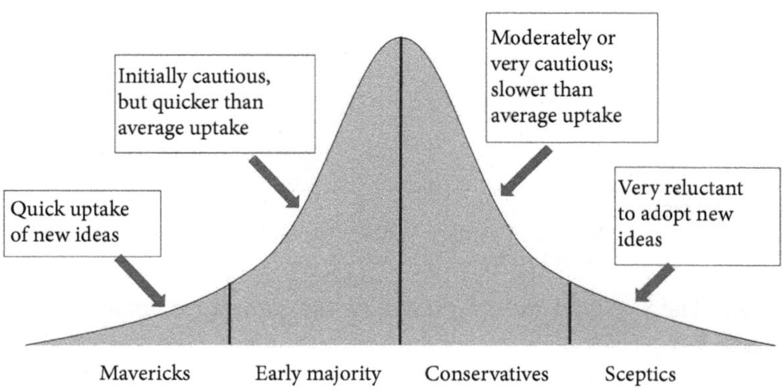

Figure 5.1 A simple model of the range of attitudes in the scientific community, and the corresponding rate of uptake of new scientific ideas

or practices.[50] In his classic work on the 'diffusion of innovations' (first published in 1962), Rogers (2003) analyses the rate of adoption of innovative ideas by members of any social system, and writes,

> When the number of individuals adopting a new idea is plotted on a cumulative frequency basis over time, the resulting distribution is an S-shaped curve [Figure 5.2]. At first, only a few individuals adopt the innovation [...]; these are the innovators. Soon the diffusion curve begins to climb, as more and more individuals adopt in each succeeding time period. Eventually, the trajectory of the rate of adoption begins to level off, as fewer and fewer individuals remain who have not yet adopted the innovation.
>
> (p. 23)

We can apply this same thinking to a scientific community consisting of individual scientists considering whether to get on board with a particular scientific idea, such as atomism; one can take the y-axis of Figure 5.2 to represent the percentage of the scientific community accepting the atomic hypothesis. For a solid scientific consensus to form, all scientists disposed to be conservative, and many scientists disposed to be outright sceptical (the

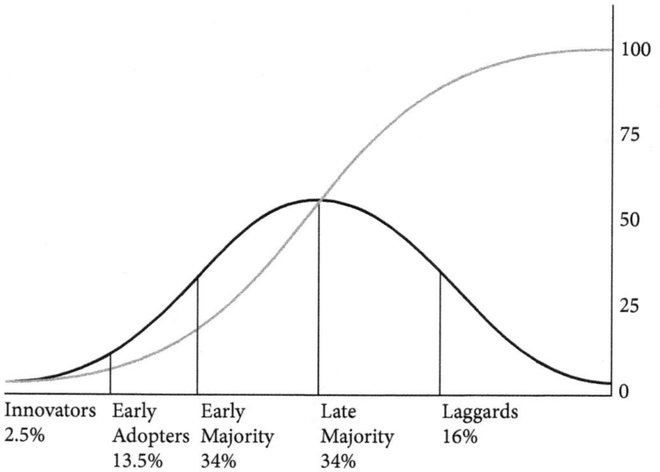

Figure 5.2 The S-shaped curve representing the cumulative uptake of an 'innovation'

[50] For a range of examples, see Rogers (2003), especially pp. 23, 46, 208–11, 268–77.

'laggards'), will either need to be convinced, or will need to die off/retire and be replaced by a new generation who are brought up as believers.

Our protagonist is trying to persuade us that the evidence really *must* have crossed the future-proof threshold in order for a solid, international scientific consensus to form. However, the hard-line conservatives and sceptics/laggards might eventually join the majority for reasons *other* than the strength of the evidence.[51] And if that's not plausible, there is the possibility that the sceptics die out, and, for sociological reasons, the next generation is educated in such a way that there is no real scope for them to doubt the scientific ideas in question. That is, one way or another, the strength of the *evidence* isn't responsible for the resulting scientific consensus.

Is this plausible? The emergence of a solid scientific consensus vis-à-vis some scientific idea, for reasons *not* having to do with 'objective evidence', is a very Kuhnian idea. Kuhn (1962) famously argued that scientists are brought up to live in a 'paradigm', an interconnected set of beliefs, theories, methods, concepts, and values. The extent of the integration is such that there is little scope for deep-level critical thinking; each aspect of the paradigm both supports, and is supported by, all the other aspects. Thus scientists living in a paradigm don't doubt the basic scientific tools they are trained to accept, and this includes scientific theories. Since there is no room for doubt, it is natural to speak of a scientific consensus. When Popper (1970, p. 52) wrote that 'The "normal" scientist, as described by Kuhn, has been badly taught. He has been taught in a dogmatic spirit: he is a victim of indoctrination,' Kuhn replied, '[S]cience students accept theories on the authority of teacher and text, not because of evidence. What alternatives have they, or what competence?' (Kuhn 2012, p. 80f.). If Kuhn is right about this we can explain the transition to a solid 95 per cent consensus of opinion vis-à-vis the atomic hypothesis *without* accepting that the evidence must have crossed some threshold such that the atomic hypothesis is definitely true. Instead we can explain it in terms of the sceptics gradually leaving the debate (e.g. retiring), and a change in the way the next generation of scientists are educated.

Kuhn's ideas have been hugely influential, and there are still many scientists, historians, and philosophers who would describe their views as 'Kuhnian'.

[51] Cf. Nitecki et al. (1978) discussing the acceptance of continental drift: '[W]e might suppose that there was at some period—perhaps during the mid-1960s—a "chain reaction" or other general shift in opinions toward the theory ["continental drift—plate tectonics"]; this more or less uniformly altered the attitude of the majority of the profession as a group and was not, at least in most cases, the result of individual judgments of the accumulating evidence and arguments for and against the theory. While the findings do not compel this interpretation, we believe that it is a plausible hypothesis that merits further testing' (p. 664).

Even Oreskes—a determined defender of science in the face of modern-day scepticism—apparently accepts the possibility of future paradigm change, even when it comes to our current best scientific theories. In Oreskes and Conway (2010) the authors fight the doubters (quite magnificently, I may add) when it comes to a range of scientific claims, including anthropogenic climate change. And yet in Oreskes (2004) we find, 'The scientific consensus might, of course, be wrong. If the history of science teaches anything, it is humility.' This is precisely what a strongly minded modern-day Kuhnian sceptic would wish to emphasise: however strong the evidence seems *now*, and however strong the scientific consensus, we should *learn from the history of science* and allow that there is a significant chance of a major change in scientific thinking in the future. Oreskes (2019) continues the point in her recent book *Why Trust Science?*, writing,

> [D]espite the claims of prominent scientists to the contrary, the contributions of science cannot be viewed as permanent. The empirical evidence gleaned from the history of science shows that scientific truths are perishable. (p. 50)[52]

And some high-profile scientists sometimes wish to make the same point. In his acceptance speech for a Career Contribution Award from the Geological Society of America, Warren B Hamilton implored,

> [...] So I appeal to all of you, as judges at all levels, from what you and others write to whom you support or hire or promote, to recognize that consensus may not define truth. [...] Changes as profound as plate tectonics, and as unanticipated by the majority, likely lie ahead.[53]

How might our believer in future-proof science reply?

She might appeal to the fact that there has been much criticism of Kuhn's philosophy of science since 1962. As already mentioned back in Chapter 1, Kuhn puts significant weight on examples from *physics*, and a certain amount of cherry-picking is evident when it comes to the historical examples Kuhn draws on to support his 'cycle of scientific revolutions' theory. Consider, for example, Godfrey-Smith (2003):

[52] I think it's clear that Oreskes would *not* wish to apply her assertion—'scientific truths are perishable'—to continental drift, nor my other examples of future-proof science listed in Chapter 1. Thus she urgently needs to clarify: *where* exactly does this assertion apply, and where doesn't it?

[53] See https://www.geosociety.org/awards/07speeches/sgt.htm for the full speech (last accessed February 2022). See Rossetter (2022) for further details on Hamilton's views.

> Kuhn seems to have been hugely influenced by the fall of the Newtonian picture of the world at the start of the twentieth century. Many philosophers of science seem to have been made permanently pessimistic about confirmation and the accumulation of factual knowledge by this episode. But Kuhn, and perhaps others, was surely *too* focused on the case of theoretical physics. [...] if we look at other parts of science—at chemistry and molecular biology, for example—it is much more reasonable to see a continuing growth (with some hiccups) in knowledge about how the world really works. We see a steady growth in knowledge about the structures of sugars, fats, proteins, and other important molecules, for example. There is no evidence that *these* kinds of results will come to be replaced, as opposed to extended, as science moves along. [...] So Kuhn's pessimism regarding the accumulation of knowledge about the structure of the world in science seems badly overstated. (p. 98, original emphasis)

This strikes me as a very reasonable criticism of Kuhn. However, there is a gap in the reasoning. One might accept that solid scientific consensuses have formed vis-à-vis the structures of sugars, fats, proteins, and so on. But to assume that this means we have 'factual knowledge' of such structures is to beg the question at issue, concerning the link between consensus and truth. Kuhn may well be wrong to highlight past scientific revolutions in theoretical physics and use these examples to make claims about science generally. But that doesn't mean he is wrong when he says that 'science students accept theories on the authority of teacher and text, not because of evidence.' It doesn't help our protagonist to establish that contemporary scientific consensuses could *only* have come about because the evidence is so incredibly strong that even most hard-line conservatives and sceptics have been won over (leaving only radical sceptics who can reasonably be ignored).

Optimistically, one might insist that a sizeable fraction of the scientific community will be sufficiently inclined towards critical thinking that they will be able to critically examine the aspects of the paradigm they have been educated to accept, including basic theoretical assumptions and theories. One may appeal to the widely held view that it is in the nature of a scientist to be sceptical, withholding belief until convinced by evidence. To illustrate:

> [M]ost good scientists are skeptics. They are disbelievers at heart.
> (Wallace and Sanders 1997, p. 21)
>
> [I]t is in the nature of scientists to be skeptical. (Feuer 2006, Chapter 3)

> [S]cientists are skeptics. They insist that any claim be supported by evidence, and they insist on a substantial debate about the quantity and quality of that evidence before accepting it. (Washington 2013, p. xi).

> By nature, scientists are a conservative and cautious bunch, especially when they have to come up with a consensus document. (Boudry 2020)

If this common conception is correct it seems implausible that scientists could believe things, not because of evidence, but rather because they have been told to believe them.

However, there are reasons to doubt the capacity for effective critical thinking within modern scientific communities, particularly when we think about accepted background assumptions and 'established' scientific results. Stanford (2019) makes the case that scientific critical thinking is squeezed out in a way it never used to be. There are a number of reasons for this, including:

(i) Science is so very interconnected these days: research teams on different sides of the world have been brought up with the same theories, methods, and basic 'background assumptions'.
(ii) Junior scientists need to attach themselves to a research project—there is little or no freedom for junior scientists to pursue their own ideas.[54] These research projects will be headed by a senior academic, who is likely to be conservative in his/her thinking, at least in many respects.
(iii) Scientific funding bodies are risk-averse, and this makes applicants pre-emptively risk-averse; they are unlikely to pursue projects challenging accepted theories, assumptions, concepts, and so on.

Stanford (2019) considers the rise of so-called 'Big Science', and comments,

> [L]arger and more collaborative projects involve steadily increasing numbers of central players and institutions, while the degree to which any given proposal departs from existing theoretical orthodoxy remains limited by the perceived chances of rejection by a review panel or funding agency that the most risk-*averse* member of that collaboration will accept.
>
> (Section 1.3, original emphasis)

[54] In the years of the continental drift controversy, for a junior scientist there was often zero freedom to believe in drift; cf. Hallam (1973, p. 105): '[S]o strong was the feeling against continental drift until quite recently that in some institutions an open adherence to this doctrine would have put at serious risk the attainment of tenure by junior faculty members.'

So there are reasons to believe that, when it comes to the freedom of thought of individual scientists, things are worse than ever before.

One can add to these considerations the fact that in past times a scientist working in a given field could personally learn almost everything he/she needed to know. These days this isn't remotely possible, and scientists must take a great deal on trust—information accumulated from books, articles, colleagues, and conferences that the individual scientist hasn't time to personally verify. For one thing, there is too much evidence, and too many new publications coming out for any individual to digest.[55] For another, evidence for a theory often spans across different disciplines, and no scientist is an expert in all of these different disciplines. Consider the theory of evolution: specialists in genetics are (usually) rank amateurs when it comes to fossil evidence, for example. Similarly for continental drift, considering diverse fields such geology and paleobotany. Indeed, it is curious to note that the scientist and the non-scientist occupy quite similar epistemic positions: *both* have to take a great deal on trust, since both lack a great deal of the requisite expertise (and, typically, the time and effort that would be required) to dig into all of the details for themselves (cf. Hardwig 1991).

Applying these concerns to the evolution consensus, a sceptic may follow Warren B Hamilton in asserting that scientific consensus doesn't equal truth, nor does it even rationally compel us to believe. Many scientists within the consensus—it will be claimed—have *not* judged (all) the evidence for themselves, but have instead trusted the judgements of their colleagues or perhaps some senior scientists in the field (at least, concerning a *lot* of the evidence).[56] Thus their names on the list of committed believers count for nothing. Indeed, the sceptic may claim, if only such scientists *did* take the time to assess the first-order evidence for themselves (e.g. they were allowed 10 years' research leave to do just that) many of them would probably change their minds, realising that the evidence is nowhere near as strong as they had originally assumed. Such a reliance on trust—sometimes *misplaced* trust—when it comes to evidence, is obvious in many cases from the history of science. In the case of

[55] Philosophers of science often assume—implicitly or explicitly—that one can digest the available evidence and make up one's mind in light of that evidence. For example, Saatsi (2020, p. 57) asks the reader to consider, 'a naturalistic philosopher who takes on board all admissible evidence'. In the same paper, Saatsi also notes in a footnote: 'At the time of writing this the physics preprint archive arXiv.org contains nearly 38,000 articles with "spin" in their title' (p. 53, fn. 11). And two papers per minute (approx.) are added to the PubMed database, day and night, without any let-up (Landhuis 2016).

[56] Hamilton himself was keen to note that 'increasingly narrow specialisation has rendered scientists unable effectively to assess evidence from areas outside their specialty', to use the words of Rossetter (2022, p.15).

continental drift, for example, junior scientists in North America trusted the judgement of senior scientists in North America who were severe critics of drift, such as George Gaylord Simpson. Another important example concerns scientists' widespread acceptance of Palmer's (1954) conclusion that peptic ulcer disease couldn't possibly be caused by bacteria. Many years later it turned out that Palmer was wrong (Zollman 2010; Šešelja and Straßer 2014). Given such examples from the history of science, the possibility of a problematic bandwagon effect—what Nitecki et al. (1978, p. 664) describe as a 'chain reaction', and Glymour (2008) calls 'sheer crowd following'—within scientific communities has to be taken seriously.

These are serious challenges to our protagonist's claim that a solid international scientific consensus could *only* emerge if the evidence base is so strong that the claims in question simply must be true. There are other possible mechanisms for the formation of such a consensus. We might do more to argue against these other mechanisms, insisting that the only *plausible* explanation of the formation of such a consensus involves incredibly strong evidence. Whilst I am sympathetic to this, there may be an easier route to establishing the desired result. The situation nicely mirrors the continental drift controversy: Wegener's generation could/should have demonstrated *that* continents were drifting without attempting to answer the thorny 'Why?' question. So too, when it comes to future-proof science, it may be possible to demonstrate *that* a solid international scientific consensus is sufficient for a claim to be future-proof, without having a (fully) convincing answer to the 'Why?' question.

4. Tackling the Threshold Problem (ii): Trust Based on Past Reliability

In many cases we've moved from 'very strong evidence' to 'it's a fact *far* beyond reasonable doubt'. For example, continental drift has made this transition. There was a time—let's say 1995 just to be safe—when one could say that a solid international scientific consensus vis-à-vis continental drift had definitely formed (Figure 5.3). At that time, if one observed the criteria for future-proof science recommended in this book, one would have called continental drift future-proof. Often it is said that continental drift was accepted in the mid-1960s, so why 1995? One reason is that the bar for future-proof science—as dictated by the two criteria given at the start of Section 3—is very high. If we ask when there was a 95 per cent consensus vis-à-vis continental

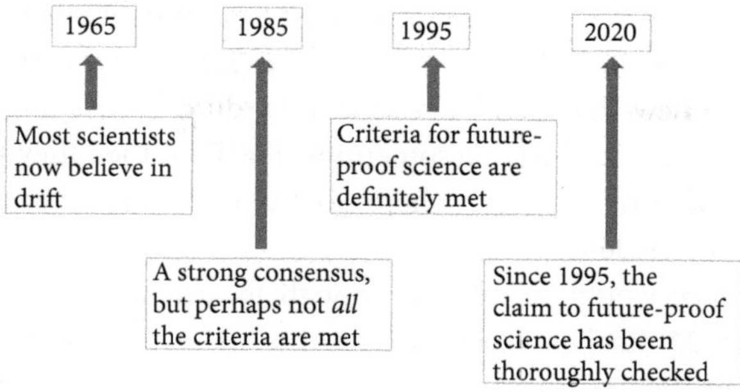

Figure 5.3 Timeline with key dates to consider vis-à-vis scientific community attitude towards drift

drift, we have to consider Nitecki et al. (1978), whose 1977 pilot study indicated that *perhaps* 11 per cent of scientists still considered the theory as 'inadequately proven' at that time. I say 'perhaps' because it was such a small study, and it is easy to pick holes in it; the study took participants from the Geological Society of America (GSA) and the American Association of Petroleum Geologists (AAPG) without worrying about whether they had a relevant PhD, and whether they had ever published anything relevant.[57] Another reason for choosing the year 1995 is that it is simply convenient to pick this year to make the point, without at all ruling out the thought that drift was probably future-proof by 1985, or 1980, or even earlier. All things considered, 1995 seems like a very safe year to choose, when we can say for sure that the two criteria for future-proof science were met.

Now we ask: Was it reasonable to label continental drift future-proof in 1995? Or was it risky? Considering the year 2020, 25 years later, we *still* had a solid international scientific consensus vis-à-vis continental drift. So in following the criteria proffered here, by 2020 you wouldn't have gone wrong (yet!) labelling continental drift future-proof in 1995. We can do much better than that, however. Consider what happened in the 25 years 1995–2020. We went from a level of evidence sufficient to establish a solid, international scientific consensus, to a level of evidence that was much, much higher. This difference is hard to conceptualise. It's the difference between massive, and much, much bigger than massive. Human beings are not good at conceptualising this kind of difference. Back in 1995 a scientist might have said (and scientists at the time *did* say), *Continental drift ought now to be simply called a*

[57] See pp. 661–2. Of the participants in Nitecki et al.'s survey, only 51% had a PhD. Another concern about the study is that they used the term 'plate tectonics' in their survey questions, but this term suggests a *theory of plate movement*, not merely the bare fact of continental drift.

'fact'. Given the amount of evidence that poured in between 1995 and 2020, how was a scientist in 2020 supposed to distinguish herself from that earlier scientist with a new statement? The evidential difference is huge, but how to articulate that effectively? Saying something like 'Back then they were sure, but now we *know*' doesn't come anywhere close to doing justice to the enormous difference in the evidence base.

The difference would be difficult enough to articulate if we were merely talking about 25 further years of research, as if the weight of evidence accumulated in the 25 years between 1970 and 1995 was matched between 1995 and 2020. However, recall from Chapter 2 the discussion of the exponential growth of science. On various different measures, science doubles in size every 20 years or so. Thus the amount of relevant science taking place in the years 1995–2020 was probably *more than double* the amount of relevant science taking place between 1970 and 1995, which was *itself* more than double the amount of relevant science taking place between 1945 and 1970. In addition, between 1995 and 2020 scientists were able to use modern instruments and techniques that were completely unavailable—even unimaginable—to earlier generations. Indeed, we have now reached a stage where it is reasonable to say that continental drift is observable. We can even observe it in different ways, which can be cross-checked with each other, confirming that the observations are ultra-reliable.[58] But even without such observations, the evidence for drift has now reached a level virtually synonymous with proof.

This case provides an illustration of how this strategy for identifying future-proof science is supposed to work. First we look back to a time when a solid international scientific consensus formed. Then we consider where we are now. The important point is this: the evidence now is such that we are 100 per cent sure—there really can be no room for doubt, unless one is a truly radical sceptic.[59] Thus the consensus that formed back then has been

[58] Continental drift is observable via Very Long Baseline Interferometry (VLBI) (see e.g. Jagoda and Rutkowska 2020), Space Laser Ranging (SLR) (see e.g. Pearlman et al. 2019), and the Global Navigation Satellite System (GNSS) (see e.g. Kierulf 2017). To illustrate, scientists have watched Maui island in the Hawaiian archipelago move approximately 1.4 metres in a north-westerly direction over the past 20 years; see https://sideshow.jpl.nasa.gov/post/series.html (last accessed February 2022).

[59] It's crucial that we don't mix up 'certainty' in the normal sense of the word, and 'certainty' in the philosopher's sense. In the philosopher's sense, I can't even be sure I am typing on a laptop right now, since an evil demon might be deceiving me. These senses of certainty do get mixed up sometimes, as when Godfrey-Smith (2021) writes, 'Nothing is certain, and history has shown that big surprises do come' (p. 312). But I'm sure Godfrey-Smith—who is no scientific sceptic—would agree that we are 'certain' about the water cycle, or that the Sun is a star. As discussed in Chapter 1, we needn't seriously entertain the thought that we might all be dreaming, or that an evil demon might currently be manipulating our thoughts.

thoroughly tested, and thoroughly vindicated. If it was going to fall, the huge quantity of relevant science that has happened since 1995 would have brought it crashing down, without a doubt. But that's not what happened. A really strict Popperian would say that the consensus view has *merely* stood up to a barrage of tests—none of this shows that continental drift is *true*. But again, given the current state of the evidence base this is borderline absurd, at least within professional scientific circles.

There are two dimensions to the modern evidence for continental drift. One is that drift is observable with ultra-sensitive modern instruments. This makes it easy to say that we know 100 per cent that drift is a fact—it is the silver bullet we were looking for in Chapters 3 and 4. Even somebody who doesn't understand much at all about how scientific evidence works can understand what it means to say that continental drift can be observed from space using GPS satellites. It eliminates any reasonable doubt in a heartbeat, even without understanding the evidence from paleobotany, paleontology, geology, and so on.[60] The *other* dimension to the modern evidence is the overwhelming quantity of 'normal' evidence, by which I simply mean scientific evidence that doesn't constitute an *observation* of drift. For example, geological, paleontological, and paleobotanical evidence that South America and Africa really are 'perfect' jigsaw pieces is vastly greater now than it was back in 1995. However, if this more-and-better evidence were all we had to go on, that might raise a concern. It's great if we can say 'Back in 1995 scientists were sure, and now we *know* 100 per cent that they were right, because drift can be observed.' We can't make this inference in quite the same way if the truth of the scientific claim in question is *not* now observable, and all we have is more-and-better 'normal' evidence, most of which the individual is not qualified to properly understand and evaluate.

But perhaps we don't need to. The core claim I want to make is that no solid international scientific consensus (in a large and diverse community) has ever been overturned. And this is no minor point, since the opportunity for such consensuses to be overturned is inconceivably huge, given (i) the exponential growth of science, and (ii) modern techniques for probing the scientific claims in question, that were completely unavailable and often even unconceived

[60] What counts as 'an observation of X', for some phenomenon X, is decided by the relevant scientific community. If we are in any doubt we can ask whether 95% of the relevant community would call it a genuine and reliable, veridical observation. For further discussion of observation in science, see Chapter 8.

when the consensus formed.[61] Those consensuses—concerning the 30 examples listed in Chapter 1—are thus incredibly solid. This point would still stand, even if *none* of the 30 cases had been proven by observation. The fact that some of them *have* been so-proven simply adds to the strength of the inference from 'solid international scientific consensus' to 'future-proof'. But more importantly, the cases where one can now observe the truth of the claim in question fight against the thought that perhaps these consensuses never get overturned because of the bandwagon effect (or some similar epistemic problem). If that *were* the case, then it would have been exposed in at least one case, when a new technique for observing something that was previously theoretical (such as drift) became available. Opportunities for this to happen within my 30 examples of future-proof science include:

(1) observations of continental drift;
(2) observations of the rotation of the Earth and the Sun;
(3) observations of the circulation of the blood;
(4) observations of the process of natural selection (e.g. White et al. 2020);
(5) observations of sound waves (e.g. Hargather et al. 2010);
(6) observations of the molecular structure of a great many different molecules, including DNA (e.g. Oteyza et al. 2013; Marini et al. 2015);
(7) observations of the structure and operation of individual biological cells (e.g. Fukuoka et al. 2014);
(8) observations of viruses (see Chapter 8);
(9) observations of the immune system (e.g. Figure 5.4).

In all of these cases there was great opportunity for the consensus view to be overturned—the theoretical ideas have been tested in the most severe way a theoretical idea *can* be tested. Since these 'tests by observation' have never led to the overturning of the corresponding theoretical ideas, an appeal to 'paradigm indoctrination', the bandwagon effect, or similar, is a thoroughly implausible explanation of the longevity of scientific consensuses generally, including those concerning theoretical ideas that have *not* been tested by observation, and perhaps *cannot* be so-tested. McMullin (1984, p. 33) touches briefly on this line of argument when he writes,

[61] The discussion in Kitcher (2019) is also helpful; for example, he stresses that, in science, 'there are large rewards for showing that some consequential piece of current orthodoxy is mistaken.'

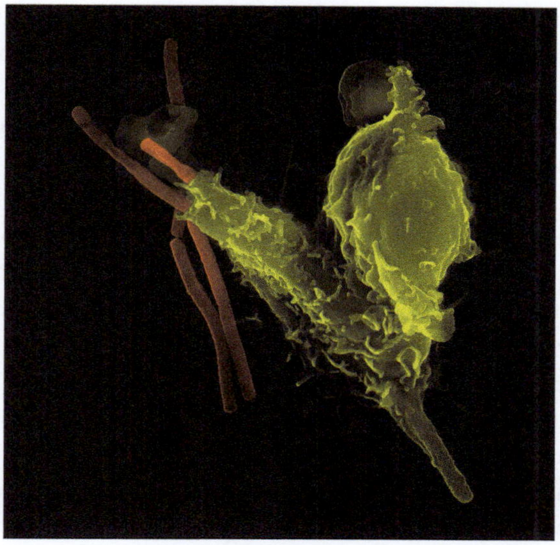

Figure 5.4 Scanning electron microscope image of a single white blood cell (yellow/right) engulfing anthrax bacteria (orange/left)

> One further point is worth stressing in regard to our geological story. Some theoretical features of the model, such as the midocean rifts, could be checked directly and their existence observationally shown. Here, as so often in science, theoretical entities previously unobserved, or in some cases even thought to be unobservable, are in fact observed and the expectations of theory are borne out, to no one's surprise.[62]

A crucial step in this book's dialectic has now emerged. The 30 examples in Chapter 1 are not merely acting as examples of future-proof science. They don't *merely* meet the criteria for future-proof science put forward earlier in this chapter. Crucially, they met those criteria *a long time ago*, and a *huge* amount of scientific work ago. In the period of time between meeting those criteria and the present day, the relevant scientific ideas have been put to the test *hugely* more than they had already been put to the test when they first met

[62] Allzén (2022) has recently explored a similar dialectic: "One of the core issues in contemporary scientific realism is how to convincingly argue for the reliability of IBE with respect to unobservables [...]. The solution, building on work by Bird, was to look more closely at cases in the history of science where explanatory inferences from predictive success to truth were made to entities whose observability was undetermined at the time, but were at a later stage found to be observable." (Section 7, 'Summary').

the criteria. In no case has this extra testing brought about a revolution in thinking. This goes a long way towards demonstrating that meeting those criteria is an exceedingly reliable—perhaps even infallible—indicator of future-proof science. And this stands even if we can't tell a convincing story as to *why* those criteria are so reliable.[63]

Before closing this section, it must be noted that *some* of what I say here could have been said by generations past. Consider Figure 5.5. If we really want to argue that we are epistemically privileged compared with past generations, then it is crucial that the way in which we epistemically distance ourselves from a past generation '2' cannot be mirrored by the way '2' could have distanced *them*selves from an even earlier generation '1'. This is by no means straightforward, since the most obvious appeal—to 'more and better evidence'—is common to all generations when they look to the past.

An important difference is the fact that several of my examples have now been vindicated by observation, essentially proving that they are future-proof, and generation 2 couldn't have said the same about their examples. But I do want an argument that all 30 of my examples are future-proof, not just the ones that have been proven by observation. The handful of examples that have been proven by observation help to show that the threshold defined by my two criteria is a reliable indicator of future-proof science. But I want to say something else. I want to say that the claim that the Sun is a star is absolutely secure, despite it not being proven by observation in the way that drift has. Why is it absolutely secure, then? I want to say: because of the weight of *normal* scientific evidence for its truth. However, this now once again sounds like the kind of thing past generations could have said about *their* claims, some of which were overturned.

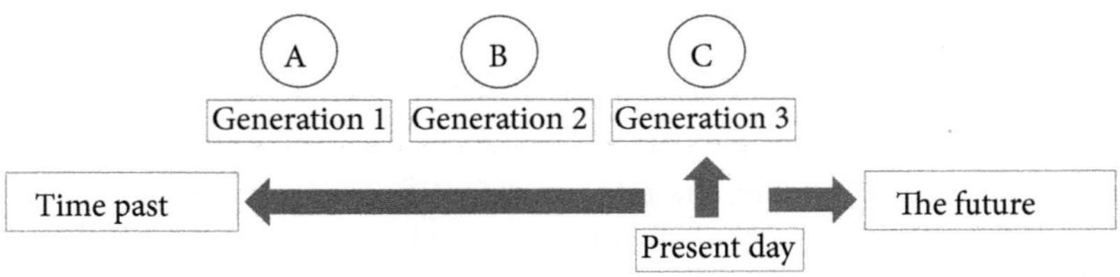

Figure 5.5 Three generations of scientists

[63] This kind of 'black box' approach is similar to that put forward by Oreskes (2019), when she answers the question 'Why trust science?' One can avoid the question as to *why* science is trustworthy, and instead opt to present evidence *that* it is trustworthy. The idea is that the latter is sufficient for rational action. Kitcher (2019) demurs.

But—to reiterate—even most 'sceptics' or 'antirealists' believe that we know the Sun is a star.[64] In admitting this, they accept that the incoming scientific evidence for this claim has gradually taken us on a journey from 'speculation' to 'theory' to 'fact'. Nor is this the kind of (weak) scientific 'fact' that we might reasonably suppose will one day be overturned (as past 'facts' sometimes were; see Chapter 8 for an alleged example). Once this is accepted, we accept that it is *not* reasonable to suppose that some future generation '4' will look back on *us* and say that we were misguided, and some even-more-future generation '5' will look back on '4' and say that *they* were misguided, and so on. Certainly new generations of scientists will come and go, probing scientific ideas to greater and greater depths—the measure of scientific scrutiny goes up and up. But when we conceptualise epistemic security, its measure can't go up and up indefinitely in the same way. There comes a point when the weight of scientific evidence for a claim is such that our degree of belief comes *very* close to 1 ('proven'). We may accept that an inconceivably huge amount of new scientific evidence may come in in the next 100 years (say), and yet insist that, since our doxastic state was already close to 1, it simply can't move much further up the scale, however much evidence comes in. That is, when we consider the relationship between the quantity of scientific evidence and our degree of belief, at some point we have very significantly diminishing returns. If the sceptic urges that in the future we may come to realise that we only *thought* we were close to 1 ('proof'), we can appeal to the fact that sceptics *themselves* don't deem that plausible, regarding a huge number of claims, both scientific (e.g. 'The Sun is a star') and non-scientific (e.g. historical claims). Sometimes—whether in the courtroom, in science, or indeed in everyday life—it is just obvious that the evidence has reached a stage where it would be wholly unreasonable to remain sceptical vis-à-vis some claim which is under scrutiny. If the sceptic will keep on insisting that perhaps the evidence just *seems* too strong for the claim in question to be wrong, at some point she is comparable with the eccentric who would caution us against believing in the Apollo Moon landings.

5. Conclusion

The goal of this chapter was to link scientific consensus with future-proof science, using the case of continental drift as a lesson from history. I've shown

[64] A determined sceptic might suggest that 'The Sun is a star' is either contentious, or is trivially true, depending on how we define 'star'. It might also be suggested that 'The Sun is a star' *could* go the same way as 'Pluto is a planet'. This actually isn't plausible, but, if preferred, another example could be used, such as 'Red blood cells carry oxygen around the body.' The specific statement 'The Sun is a star' will be analysed in Chapter 9, in 'Objections and Replies', Section 2.4.

that continental permanency ('fixism') was never 'settled science' in the way those 30 examples from Chapter 1 are. Thus it is relatively easy to draw a line between cases such as continental permanency—where most scientists are (strongly) in favour but a significant number are not and debate is ongoing—and other cases where scientists are overwhelmingly certain and debate has been closed down. A worry arose that the line I have drawn is ad hoc, chosen purely to render those historical challenges mute, and preserve my favoured conclusion that most current scientific theories are true. I responded in two ways. First, there is reason to suppose that in many cases we have now crossed a threshold, where the evidence is so strong that it would be wholly unreasonable to remain sceptical. Second, some of my cases have now been essentially proven future-proof, because what was once theoretical is now observable. These 'proven by observation' cases help to demonstrate that the threshold I have chosen—defined by the two criteria stated in Section 3 of this chapter—is sufficient for future-proof science generally. If it were not, then it is highly likely that the scientific claim in question would have been overturned in at least one case by now. In fact, no case meeting my criteria has *ever* been subsequently overturned. And it is implausible that this can be explained by the fact that scientists are all stuck in a paradigm, especially in cases where what was once theoretical is now observable.[65]

But perhaps it could be objected that there *are* cases from history where my criteria were met, and then subsequently the scientific ideas in question were overturned. There was once a solid consensus that the Earth sits at the centre of the universe, with everything orbiting around it. This was a long time ago, however, and it is easy to dismiss it as 'not proper science'; the contrast between 16th-century practices and modern practices is stark. A better example concerns the confidence in the truth of classical physics during the 18th and 19th centuries. Whilst the *equations* of classical physics remain very useful and are accurate in a great many contexts, the underlying physical ideas have changed dramatically (cf. Stanford 2006, pp. 182–3). Indeed, the history of physics shows time and time again how equations can achieve a great deal, even when the preferred physical interpretation of those equations is—with the benefit of hindsight—way off the mark. It is no surprise that Kuhn put most dialectical stress on examples from physics. But equally, the case of

[65] The *directness* of observation is a sliding scale: light microscopy is more direct than electron microscopy, which in turn is more direct than the observation of continental drift using GPS satellites. For discussion, see Chapter 8.

physics should not be considered paradigmatic of science generally, even if physics has historically often been considered the 'premier' field of science.

Thus we need to take seriously the thought that non-classical theoretical physics, or perhaps 'fundamental physics' (broadly construed), is a special case when it comes to the question of future-proof science. This we turn to next.

6

Fundamental Physics and the Special Vulnerability to Underdetermination

1. Introduction

We've reached an important point in the overall dialectic, and it's a good time to take stock.

I've been looking for a way to identify future-proof science reliably, perhaps even infallibly. In Chapters 3 and 4 I gave up on identifying it by directly digesting the first-order evidence. An alternative approach is to identify it via a solid scientific consensus. The word 'consensus' is used in various ways—with various strengths—in the literature, so I needed to pin down precisely what I mean by 'solid consensus', such that we can be sure the corresponding scientific ideas are future-proof. In Chapter 5, Section 3, I introduced two criteria which, when satisfied, certainly give us a scientific consensus that is 'solid', and which, I argued, are collectively a sure sign that the relevant scientific ideas are future-proof. Ideally I would compare scientific ideas *claimed* to be future-proof with scientific ideas that actually *are* future-proof, showing that these match up. It has often been said by sceptics that this sort of move is clearly impossible. It is like the age-old issue of trying to compare a representation of reality with reality itself, without ever having access to reality *except via a representation.*[1] In the current context, what is clearly impossible—so the sceptics say—is for us to identify the scientific ideas that actually *are* future-proof: that would require us to see (very far!) into the future. We can no more do this than we can lift the veil of perception.

However, I found a way around this problem. Sometimes scientific claims, at one time based on theory and experiment, at a later time come to be testable by observation. The case of continental drift is a pertinent case, as discussed in the previous chapter. The solid consensus vis-à-vis the truth of continental

[1] Cf. Jordan (2013, p. 197): '[I]t is logically impossible to compare experience (i.e., representations) to nonexperience (i.e., nonrepresentations). [. . .] [I]f consciousness is an internal representation, then all that consciousness can ever compare is its own representations to its own representations.'

Identifying Future-Proof Science. Peter Vickers, Oxford University Press. © Peter Vickers 2023.
DOI: 10.1093/oso/9780192862730.003.0006

drift that formed in the latter half of the 20th century can be compared with the fact that we nowadays have amazing instruments that can actually *observe the drift happening*. For example, the island of Maui in the Hawaiian archipelago has moved approximately 1.4 metres in a north-westerly direction over the past 20 years.[2] Such observations show—beyond all reasonable doubt—that 'drift theory' really is future-proof, and we don't need to see into the future to know that. Thus the impossible comparison is *not* impossible: we can compare the solid scientific consensus vis-à-vis drift that formed in the second half of the 20th century with our present-day knowledge that drift actually is future-proof.[3]

I took this further to make a general point. The bar I have set for future-proof science seems sufficiently high, since it has been thoroughly tested time and time again, in a great many different contexts. In all 30 of the examples of future-proof science listed in Chapter 1, since a solid consensus formed there has been *so much* opportunity for later science to prove the idea wrong that it is hard to conceptualise. This is owing to both the huge amount of relevant scientific work that has taken place as science continues to grow exponentially, and also the invention of new techniques and instruments that are able to probe the scientific claims in ways not even imagined when the consensus originally formed. In this way we reach our conclusion, that a scientific claim meeting those two criteria is future-proof. Since there isn't a single counter-example to this in the entire history of science, my claim is *not* intended to be defeasible. We really should believe that those 30 examples given in Chapter 1 will last forever, not merely 'most' or 'many' of them.[4]

But one rather significant concern may be raised. It starts with the point that one cannot generalise much when it comes to science. Different scientific fields work in significantly different ways, with different methods and with different types of evidence taking priority (as discussed in Chapter 2). Such a warning

[2] See https://sideshow.jpl.nasa.gov/post/series.html (last accessed February 2022).

[3] Obviously we cannot observe drift happening 50 years ago (say); some extra assumptions are required to infer that drift was happening back then. However, it would be highly unreasonable to suppose that drift might have started just before we started observing it. Our observation of drift happening now serves to confirm what scientists were already sure about back in the 1990s—that drift has been happening for many hundreds of thousands of years.

[4] A natural fallback position for me, should there be any doubt regarding my arguments, is to more modestly assert that any scientific claims meeting the stated criteria are *almost certainly* future-proof. Compare the strategy in Miller (2013), where we find, '[M]y account does not *guarantee* that in all cases where my conditions obtain the consensus is knowledge based, only that it is *likely* to be so' (p. 1300, original emphasis). Those readers with a natural inclination to be extremely cautious when it comes to matters epistemic will prefer to include the 'almost certainly' qualification (and it would be naïve to think that philosophical arguments have the power to render natural inclinations irrelevant). But even the more modest assertion entails that doubt regarding any one of my examples, taken individually, would be unreasonable.

against generalisations about 'science' needs to be applied to the question of whether the given criteria are sufficient for future-proof science across *all* scientific fields. When it comes to the relationship between theory, evidence, and truth, one field of science is particularly noteworthy: the case of non-classical theoretical physics, or 'fundamental physics' (broadly construed). Importantly, there are no cases of 'fundamental physics'—a term to be scrutinised in due course—in my list of 30 examples of future-proof science. So if fundamental physics is indeed a special case, then those 30 examples should do little to persuade us that a claim *from this scientific context* meeting my criteria will be future-proof.

Some stripe or other of non-classical theoretical/fundamental physics has often been singled out as a special case. In his 'A Case for Scientific Realism', McMullin (1984) explicitly contrasts quantum mechanics (QM) with examples of scientific knowledge from evolutionary history, geology, molecular chemistry, and cell biology, noting that '[m]uch depends on the sort of theoretical entity one is dealing with' (p. 30), and 'The antirealist cannot, it seems to me, make sense of such sequences [theoretical continuities as science advances], which are pretty numerous in the recent history of all the natural sciences, *basic mechanics, as always, constituting a special case*' (p. 33, my emphasis). Godfrey-Smith (2003, p. 98) draws on McMullin, writing that 'It is possible that, when we try to work out how to describe the growth of knowledge over time in science, we should treat theoretical physics as a special case and [contra Kuhn] *not* as a model for all science' (original emphasis; cf. Godfrey-Smith 2021, p. 228). Fahrbach (2011, p. 151, fn. 14), providing his list of 'true' scientific theories, writes, 'Note that my list does not contain any theories from fundamental physics. Such theories have special problems which need special treatment.' Recently, Hoefer (2020, p. 19) has explicitly argued that '[realism] should exclude the most fundamental physical theories we have right now'. Stanford's problem of unconceived alternatives is also highly relevant, as already noted in Chapter 2: it seems reasonable to say that there are special problems conceiving plausible alternatives in the context of fundamental physics. What counts as a 'plausible' alternative has sometimes been highly controversial in this particular context of scientific theorising.

The purpose of this chapter is to take a stance on 'fundamental physics' (broadly construed), and in particular the worry that my future-proof filter might *not* be trustworthy in this context, even if it is trustworthy in all other scientific contexts. In the next section the worry is brought to life via a discussion of Sommerfeld's 1916 theory of the hydrogen fine structure. In Section 3, I consider a range of different ways of drawing a principled epistemic distinction, and find problems with them all. In Section 4, I more carefully examine the case for doubting the future-proof filter in this context, and find it wanting—I argue that the filter can be trusted in this context after all. The only caveat is that sometimes claims coming out of fundamental physics require careful interpretation, since they often don't wear their meaning on their face. These problems of interpretation are examined in Section 5; Section 6 concludes.

2. The Sommerfeld Miracle

As already noted in previous chapters, over the past 50 years philosophers searching for hallmarks of truth in science gradually focused more and more on novel predictive success. At the same time, some scholars were noting why we must be highly cautious about truth claims when it comes to theoretical physics. And yet, modern theoretical physics is the source of the most startling novel predictive successes in the entire history of science! This is a curious juxtaposition; should we be cautious about theoretical physics, or confident?

The extraordinary predictive success of quantum electrodynamics (QED) is often noted. Worrall (2007, p. 125) writes,

> Quantum Electrodynamics predicts the magnetic moment of the electron to a level of precision better than 1 part in a billion. How, it seems natural to ask, could a theory make a prediction about what can be observed that turns out to be correct to such an amazing degree of accuracy, if what it claims is going on 'behind' the phenomena, at the level of the universe's 'deep structure', is not itself at least approximately correct? This may be logically possible, but it seems none the less monumentally implausible.

However, consider this same quotation, but now substituting references to QED with references to Sommerfeld's 1916 theory of the hydrogen atom, and making the prediction in question Sommerfeld's prediction of the hydrogen 'fine structure' spectral lines:

> *Sommerfeld's 1916 theory of the hydrogen atom* predicts the *hydrogen 'fine structure' spectral lines* to a level of precision better than 1 part in a billion. How, it seems natural to ask, could a theory make a prediction about what can be observed that turns out to be correct to such an amazing degree of accuracy, if what it claims is going on 'behind' the phenomena, at the level of

> the universe's 'deep structure', is not itself at least approximately correct? This may be logically possible, but it seems none the less monumentally implausible.

Sommerfeld's theory has now been thoroughly replaced with modern QM, and cannot possibly be described as 'approximately correct'. So what Worrall describes here as 'monumentally implausible' actually *turned out to be the case* when it comes to Sommerfeld's theory. It is, on the face of it, just the kind of 'miracle' that is supposed to be impossible according to the scientific realist's 'no miracles' argument. The story of the Sommerfeld miracle is thus an exceedingly important lesson from history.[5]

Sommerfeld's 1916 theory built on Bohr's famous 1913 theory of the (hydrogen) atom. Bohr had posited that the hydrogen atom consists of a central, positively charged nucleus, and a single, negatively charged electron which orbits that nucleus. The electron can only occupy certain 'allowed' orbits, and it makes jumps between the orbits when it gains or loses energy. The size of the jump corresponds to the energy change in the electron. And whenever the electron *loses* energy it emits that energy in the form of light. The colour of the light emitted depends upon the amount of energy lost (the size of the jump). In this way only certain colours are emitted, and thus Bohr managed to provide an explanation for the characteristic spectral lines of hydrogen (Figure 6.1). Bohr didn't *predict* the lines, but he soon applied his theory to helium atoms that have lost one electron (ionised helium), and then his theory could make a genuine, testable prediction. The predictions worked out beautifully, and brought Einstein to remark, 'This is a tremendous result. The theory of Bohr must then be right' (in Pais 1991, p. 154). As Pais (p. 149) writes, 'Up to that time no one had ever produced anything like it in the realm of spectroscopy, agreement between theory and experiment to five significant figures.'

However, Bohr's theory was always considered a first approximation; it needed to be de-idealised. In addition, it was known at the time that the very phenomenon Bohr was famous for explaining was, in a sense, an illusion. When one looks closely at those spectral lines—with a high resolution spectroscope—one sees that they are in fact made up of a number of separate

[5] Worrall might quibble, insisting that at the level of 'deep structure' Sommerfeld's theory was approximately correct. More on this possibility in due course. Note that in the quote Worrall conflates 'what is going on behind the phenomena' with 'deep structure', and these are very different things: we don't need to go all the way to 'deep structure' when we venture 'behind the phenomena'. Sliding from one to the other is dangerous.

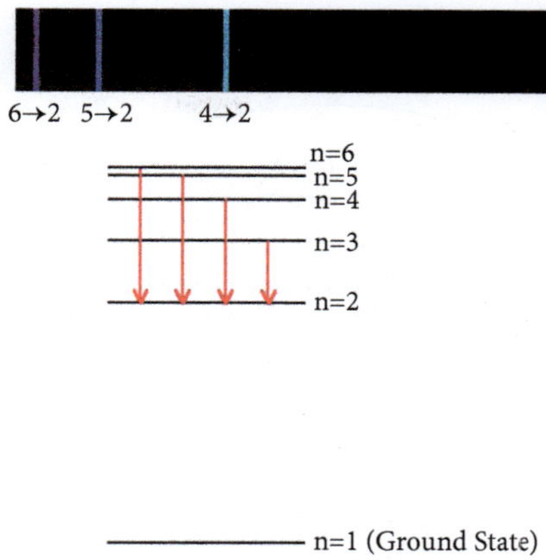

Figure 6.1 Purple, blue, and red lines emitted by hydrogen, and explained by Bohr's 1913 theory in terms of certain 'allowed' electron jumps between different values of 'n', corresponding to different possible orbits

lines. This is the fine structure. Thus a remarkable test of Bohr's ideas lay ahead. The question was posed: Was it possible to *naturally* de-idealise Bohr's theory, and *by so-doing* recover the exact phenomena that needed to be accommodated? Of course, there could be no question of adjusting Bohr's theory ad hocly, in order to recover the phenomena; the community would instantly see through such a move. Any whiff of ad hocery would be jumped on ferociously, especially by those in the community deeply opposed to Bohr's peculiar new ideas.[6] The de-idealisation would have to be completely natural, precisely the de-idealisation one was scientifically obliged to attempt anyway, even if there were no phenomena waiting to be accommodated by theoretical adjustments. If the theory really were getting at the truth of reality, this would *have* to work out. A more severe test of Bohr's ideas couldn't be imagined.

This was the challenge Sommerfeld grappled with during 1915–16.[7] Along with Wilson and Ishiwara (see Heilbron and Kuhn 1969, p. 280), Sommerfeld generalised Bohr's quantum condition—the condition dictating which orbits are 'allowed'. He did this in such a way that it unified the ideas of Bohr and Planck, and this served to block any suggestion that it was an ad hoc move. With the new 'phase integral' quantum conditions (Vickers 2020a, Section 4) both degrees of freedom for elliptical orbits were now quantised: (i) overall size

[6] For example, Ehrenfest wrote a letter to Lorentz dated 25 August 1913, stating, 'Bohr's work on the quantum theory of the Balmer formula (in the *Philosophical Magazine*) has driven me to despair. [...] If this is the way to reach the goal I must give up doing physics' (Longair 2013, p. 113f.).

[7] Bohr had already tried, without success, to de-idealise his own model. See Nisio (1973), p. 54ff.

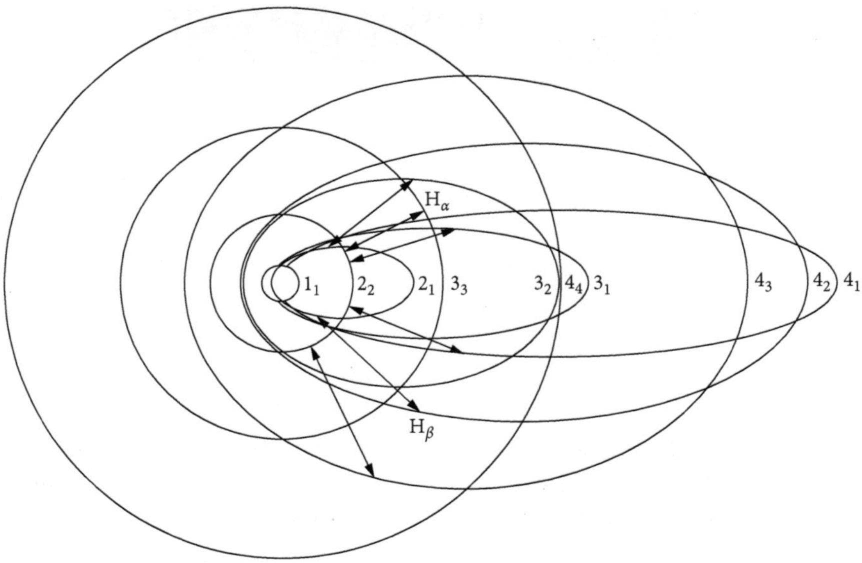

Figure 6.2 Some of the allowed electron orbits in a hydrogen atom

of orbit, and (ii) eccentricity of orbit. Bohr provided a useful visualisation of the resultant 'allowed' electron orbits in his Nobel lecture of 1922 (Figure 6.2).

It remained to now de-idealise further, by introducing the effects of relativity. There was only one way to do this, and so the question of ad hoc adjustments couldn't possibly arise. One had only to work one's way through the mathematics. Doing so, Sommerfeld derived the fine structure formula for the allowed energy states of unperturbed hydrogen, and thus via $\Delta E = h\nu$ the possible frequencies of the hydrogen fine structure spectral lines (for Z=1):[8]

$$E_{n_r,n_\varphi} = m_0 c^2 \left[1 + \frac{\alpha^2 Z^2}{\left[n_r + \left({n_\varphi}^2 - \alpha^2 Z^2 \right)^{1/2} \right]} \right]^{-1/2} \qquad [1]$$

This formula (combined with $\Delta E = h\nu$) encodes countless novel predictions of spectral lines. It even applies to different elements (not only hydrogen), so long as the atoms are ionised such that they are one-electron atoms. In particular, the formula applies to the ionised helium fine structure (Z=2).

Sommerfeld didn't know this at the time, but [1] would later be shown to be 'perfect', since it is identical to the formula that later resulted from the fully

[8] In this equation m_0 is the rest mass of the electron, c is the speed of light, α is the fine structure constant equal to $e^2/\hbar c$, n_r and n_φ are the radial and angular quantum numbers, and Z is the proton number. By 'unperturbed' I mean 'not affected by electric or magnetic fields'.

relativistic Dirac QM of 1928 (itself an improvement on 1925–6 Schrödinger–Heisenberg QM). Thus it *is* reasonable to compare Sommerfeld's fine structure predictions with the 'one part in a billion' QED prediction that impressed Worrall (2007). And yet, at the very heart of Sommerfeld's theory are continuous worldline elliptical orbits of electrons, derived using relativistic classical mechanics. Sommerfeld assumed that the mass of the electron changes as its velocity changes during its orbit, in line with relativity. But as Griffiths (2004, p. 16) notes in his popular textbook: 'It's not even clear what velocity *means* in [modern] QM' (original emphasis). That is to say, what Sommerfeld's theory 'claims is going on "behind" the phenomena' (Worrall 2007, p. 125) is definitely not at least approximately correct, however monumentally implausible that may seem in light of the empirical success.

The relationship between Sommerfeld's predictions and the empirical phenomena was a matter of *some* controversy at the time, since it was difficult to achieve the required empirical accuracy with the instruments of the day. Certainly it was clear that the match between theory and experiment was good, but it wasn't immediately obvious quite *how* good the match was.[9] Kragh (1985) analyses the community reaction of the day, writing that mainstream physicists 'were completely satisfied with Sommerfeld's theory of the hydrogen spectrum' (p. 102), concluding that 'Sommerfeld, Bohr, and their disciples decided that Paschen's confirmation of the theory was so decisive that no counter-evidence could qualify as serious anomalies' (p. 84). Paschen was the main experimentalist who set out to confirm Sommerfeld's prediction; he wrote to Sommerfeld in 1916, 'My measurements are now finished and they agree everywhere most beautifully with your fine structures' (p. 75). As Kragh continues,

> Sommerfeld's theory was generally considered to be excellently confirmed by experiments... To many physicists the theory was the final proof of the soundness of Bohr's quantum theory of the atom. For example, in letters of 1916 Einstein called Sommerfeld's theory "a revelation." "Your investigation of the spectra belongs among my most beautiful experiences in physics. Only through it do Bohr's ideas become completely convincing." Paul Epstein was converted to Bohr's theory only "after Sommerfeld in his theory

[9] An important caveat here: scientists weren't clear on which electron jumps were and weren't possible. It appeared that not *all* jumps were possible, since some lines, allowed by Sommerfeld's theory, appeared to be missing (see Vickers 2020a, Section 2 for discussion). However, the lines that *did* appear were predicted by the theory with extremely high accuracy.

> of the fine structure of the hydrogen lines achieved such a striking agreement with the experiment."
>
> During the reign of the old quantum theory, that is, until 1926, Sommerfeld's explanation of the fine structure was regarded as undisputable by the leading atomic physicists. (1985, p. 80)

And yet, Sommerfeld's theory was destined to be thoroughly replaced (not merely modified). The elliptical orbits would completely disappear, along with the classical physics that had motivated them.

In the past 100 years, scientists have often looked back on this episode and asked themselves: What exactly happened there? How was it possible for such a thoroughly mistaken theory to be so empirically successful? It has seemed to some physicists like an unfair trick—scientists hope very much, and expect, that if their theoretical ideas are wide of the mark, that will show up clearly when they are empirically tested. For science to work effectively, it is crucial that dead-end ideas are exposed when they are brought up against observation and experiment. Accordingly, in his attempt to resolve the 'Sommerfeld Puzzle', Biedenharn (1983, p. 14) describes the case as 'a sort of cosmic joke at the expense of serious minded physicists'. In fact, scientists—completely unaware of the 'no miracles argument' of the scientific realist—have sometimes described Sommerfeld's predictive success as a 'miracle', given how mistaken his ideas were. This rather makes a joke of the 'no miracles' argument, at least in the context of theoretical physics. Back in 1968 Jack Smart famously put forward his own version of the 'no miracles argument', writing, 'if we interpret a theory in the realist way, then we have no need for such a cosmic coincidence' (Smart 1968, p. 39). Smart could have had no idea that that very same year, Heisenberg (1968, p. 534) was describing the Sommerfeld prediction as just the kind of 'miracle' that was claimed—by realists such as Smart—to be too implausible to be taken seriously.[10]

Let's get clearer on what exactly the 'Sommerfeld puzzle' is. Kragh (1985, p. 84) writes as follows:

> By some sort of historical magic, Sommerfeld managed in 1916 to get the correct formula from what turned out to be an utterly inadequate model...

[10] Granovski (2004, p. 524), commenting on the Sommerfeld case, writes, 'At this point, a small miracle occurred.' Kronig (1960, p. 9) describes it as 'perhaps the most remarkable numerical coincidence in the history of physics'.

> [This] illustrates the well-known fact that incorrect physical theories may well lead to correct formulae and predictions.

Note, however, that realists are completely on board with Kragh's claim that incorrect physical theories lead to correct formulae and predictions. The real puzzle in this case is that we have *extremely risky* quantitative predictions, coming from a theory where even the most *centrally working parts* are not even *approximately* true. This can be seen by consideration of the conceptual gulf between Sommerfeld's 1916 theory and the 1928 Dirac theory. Sommerfeld used classical mechanics to construct his continuous worldline trajectories for the orbiting electron. When he adjusted the orbits using relativity, he ended up with precessing elliptical orbits. Not only was this picture incorrect, but in the 1928 Dirac theory there are no orbits *at all*. In addition, Dirac's theory includes the special new quantum property of electron *spin*, which (seemingly) plays the crucial role in Dirac's theory vis-à-vis the fine structure spectral lines. As Letokhov and Johansson (2009, p. 37) put it, 'The interaction between the spin and the electron's orbit is called spin-orbit interaction, which contributes energy and causes the fine-structure splitting.' This is the natural way to talk in modern-day atomic physics: electron spin causes the fine structure splitting.[11] Thus it makes sense to say that the theoretical elements doing the essential work, to generate the fine structure formula [1], are fundamentally different in the old Sommerfeld theory and the successor Dirac theory: on the one hand we have relativistic, precessing elliptical orbits, and on the other hand we have electron spin. The realist cannot appeal to theoretical overlap, unless that overlap is somehow at a very deep level of theoretical abstraction. For present purposes it isn't important to understand what happens in the 1928 Dirac derivation but—just to quickly summarise—it begins with the Hamiltonian $H = \rho_1 \boldsymbol{\sigma} \cdot \boldsymbol{p}c + \rho_3 m_0 c^2 - \frac{Ze^2}{r}$, where σ is the spin operator, and ρ_1 and ρ_3 are 4x4 matrices. One then solves the eigenvalue problem $H\psi = E\psi$, where ψ is a four component spinor (see, for example, Biedenharn 1983, p. 25ff.). Despite radically new ideas, and even a new mathematical language, the result of the derivation is the exact same fine structure formula, equation [1], that Sommerfeld had already derived in 1916, and that can*not* be achieved with 1925–6 non-relativistic quantum mechanics.

[11] Letokhov and Johansson (2009) do refer to 'electron orbits' here, but it is a *façon de parler*. In modern quantum physics there are no electron orbits on any interpretation of quantum mechanics, even on the Bohmian interpretation. See Section 5 for further discussion.

Of course, over the course of Chapters 3 and 4, I dismissed successful predictions as the way to identify future-proof science. Thus I might dismiss the Sommerfeld case too, stating that the successful predictions encoded by [1]—however 'perfect' quantitatively speaking—shouldn't *by themselves* persuade anyone to believe the theory, nor even its working parts, to be even approximately true. In addition, given the criteria for future-proof science stated in Chapter 5, Sommerfeld's theory doesn't come out as future-proof, most obviously because there wasn't a 95 per cent solid scientific consensus regarding it. Indeed, the period of 'old quantum theory' lasting from roughly 1913 to 1926 was a period of great upheaval, where many in the scientific community—and certainly more than 5 per cent of that community—didn't know quite what to think (Vickers 2020a, Section 2). So the reign of Sommerfeld's theory, from roughly 1916 to 1926, was never sufficient in its influence for it to be reasonable to describe it as 'future-proof'. And one might add that the relevant scientific community wasn't at all diverse, at least along some important dimensions of diversity, such as gender.[12] Thus without a careful case study it is already obvious that this case doesn't meet my criteria.

But the point of this chapter is not merely to show that even the 'ultimate historical challenge to scientific realism' (Vickers 2020a) does not speak against my criteria for future-proof science. That would be too easy, since I have raised the bar very high with my demanding criteria. What's really important about this case is to understand just *how* Sommerfeld's radically false theoretical ideas were able to lead to the 'perfect' empirical formula. It is possible to see this, in hindsight, via an analysis of the relationship between Sommerfeld's theory and Dirac's theory. The short story is that a mathematical relationship between Sommerfeld's theory and Dirac's theory is possible to uncover, in hindsight, with some ingenuity (see Vickers 2020a for the details). The highly abstract nature of the relationship serves to demonstrate how, in

[12] I think a dramatic gender imbalance (e.g. 99% men) in a community is highly significant, highly *problematic*, for epistemology. But consider the argument: all men are different, and a purely male community *can* include within it a highly diverse range of perspectives. And a highly diverse range of perspectives is all we need to eliminate problematic biases. Thus, a dramatic gender imbalance in a community doesn't *entail* an epistemic problem, and we shouldn't *insist* on a reasonable gender balance in the relevant community when we formulate criteria for future-proof science (cf. the discussion in Miller 2013, p. 1312f.). But I don't need to refute this argument: recall that my filter is supposed to be *sufficient* for future-proof science, not necessary. I will insist on a reasonable measure of gender balance, and remain silent on the question of whether a 99% male community could ever comprise such a diversity of perspectives (along e.g. dimensions of background, specialism, nationality, religion, political leaning) that a solid consensus within it ought to be fully compelling. My filter would simply knock back any claims made by a 99% male community, without passing comment on them. We needn't worry too much about this, since (I argue) all contemporary scientific communities are sufficiently diverse for future-proof science; see Chapter 9, Section 2.2, for further discussion.

theoretical physics, mathematical structure can do a lot of the work to generate predictions, whilst at the same time being compatible with radically opposed physical interpretations. This is especially obvious when mathematical relationships can be deeply hidden, and only come to the surface when manipulations are employed which force them to the surface. Such is the case with the Sommerfeld–Dirac case. The very last sentence of Biedenharn (1983) is: 'But just think how utterly hopeless it would have been to try to understand Sommerfeld's success without the detailed enlightenment afforded by Dirac's equation!' (p. 32). One obviously couldn't mathematically relate Sommerfeld's theory to Dirac's theory without already having Dirac's theory 'in hand'.

Thus a take-home lesson from this case—and others like it within the field of fundamental physics—is that Stanford's problem of unconceived alternatives bites particularly hard in this context. And recent work by Dardashti (2019, 2021) further strengthens the problem of unconceived alternatives in the context of highly mathematical, theoretical physics. Section 9.3 of Dardashti (2019) is entitled 'Problems of Theory Space Assessment', and he considers 'specific cases from the history of physics in which scientists have mistakenly constrained theory space' (p. 159). To give just one example, he notes that 'The three-dimensionality of space, a seemingly obvious empirical fact about nature, turned out to be an incredibly flexible element in theory development' (p. 165). And in Dardashti (2021) it is argued that 'no-go theorems' have time and again shown themselves to be not as 'no-go' as they originally appeared. In fundamental physics, what at one time seemed impossible does not seem so impossible at a later time, or from the perspective of a later scientific community.

3. In Search of a Principled Epistemic Distinction

One possible conclusion we might draw from the previous section is that fundamental physics is indeed a special case. One thing that makes it special is the driving force of the mathematics, which can lead to empirical success even when the physical ideas are wide of the mark. The Sommerfeld case also shows that the physical ideas, whilst being wrong, can yet be highly motivated (given background assumptions). An additional worry comes with the fact—also illustrated by the Sommerfeld case—that common sense may go out the window in the context of fundamental physics. As Bird (2007) writes, 'There is no reason why the fundamental nature of the universe should be even comprehensible, let alone intuitive' (p. 214f.). Thus we must employ extreme caution when it comes to mathematics-rich theoretical/fundamental physics, not necessarily concerning our trust in the *predictions*, but concerning the story of what is going on 'behind the phenomena'.

If our priority is epistemic caution—and in particular if we want to make the future-proof filter infallible—then we may want to simply block all claims coming out of fundamental physics. The filter would then be a *double* filter, blocking all claims from fundamental physics in an initial stage, and then employing the Chapter 5 criteria in a second stage. Crucially, this wouldn't be to say that all claims coming out of fundamental physics are *not* future-proof. The filter is designed such that all claims getting through *are* future-proof, and it is silent on claims that do *not* get through. It might seem a shame to block all such claims, especially if we think some such claims are indeed future-proof. But this might seem like the *safest* way forward, given the special considerations that accompany 'fundamental physics'.

Hoefer (2020) takes precisely this approach in his discussion of scientific realism. For Hoefer, 'whatever the exact scope of the realist's claim may be, it should exclude the most fundamental physical theories we have right now. By this I mean…', and then he gives a list of examples (p. 19) consisting of quantum theories and general relativity. One might be a realist about viruses, molecules, and all manner of other things (and processes), but fundamental physics should be off limits. For Hoefer, when it comes to fundamental physics, there is 'a special vulnerability to underdetermination' (2020, p. 21). And what *is* 'fundamental physics', for Hoefer? Crucially, he doesn't mean to refer to scientific claims concerning the ultimate laws and constituents of reality. In this sense of 'fundamental', QM is *not* fundamental, since it is 'effective', in-principle derivable from fully Lorentz-invariant quantum field theories (cf. Hoefer 2020, p. 28).[13] So too, we expect General Relativity to be in-principle derivable from some future theory of Quantum Gravity. In fact, we can't be sure that *anything in physics* is fundamental in this (strong) sense, since we are not at the end of physics. Thus this sense isn't particularly helpful for current purposes, and it certainly isn't the (weak) sense of 'fundamental' Hoefer has in mind when he writes that realism 'should exclude the most fundamental physical theories we have right now'. Indeed, if he *did* mean it in that sense he wouldn't be excluding very much!

[13] Cf. Egg (2021): 'Although QM is sometimes loosely classified as part of "fundamental physics', this can only mean that QM is fundamental *relative to* other parts of physics, for example condensed matter physics" (original emphasis).

We may ask: What positive definition does Hoefer provide for the other (weak) sense of 'fundamental', the one he does wish to put weight on when he defines the scope of his realist commitments? Clearly it is more inclusive, but how should we characterise its limits? On this Hoefer is unfortunately quite vague, mostly referring to 'quantum physics' but also including General Relativity as an example of what we should rule out. Since we already have special reasons to be epistemically cautious about quantum physics and General Relativity (in their current form), there is a danger of the distinction being unprincipled, underpinned by a list of cherry-picked examples. Hoefer does refer, briefly, to theories that involve entities and properties that are 'severely unobservable' (2020, p. 21), suggesting a distinction based on (un)observability, though at the same time I'm certain he doesn't wish to follow Van Fraassen (1980). At other times the distinction seems to be in terms of what we can/cannot, *imagine* being overturned, or what we can/ cannot *conceive* might turn out to be wrong (Heofer 2020, pp. 23–4). These are sketchy thoughts, I'm sure the author would agree.[14]

But suppose we stop asking for a definition, and instead take an 'I know it when I see it' attitude to fundamental physics. We can use 'fundamental physics (broadly construed)', to make sure that we block the interpretations of QM (say), and not only the *most* fundamental physical theories. One problem now is that we end up blocking claims we really didn't want to block. Hoefer himself is keen to include 'basic nuclear physics' (2020, p. 30) as one example of solid scientific knowledge. And I agree that there are very good reasons to include 'basic nuclear physics' in any list of scientific facts. First, we have a consensus: scientific experts don't doubt for a moment that we know the *basic*, broad-brush story of at least *some* aspects of nuclear physics. We know, for example, certain basic facts concerning the composition of atomic nuclei, nuclear stability/instability, and nuclear fission/fusion. Indeed, these things were already known many decades ago, and some of them were used to successfully construct the first nuclear weapons. Since then, our knowledge of such things has only grown and strengthened, over many decades. As this knowledge has strengthened and deepened, the community has accumulated an abundance of what Hoefer refers to as 'epistemic handles', where '[e]pistemic handles include things like: direct or indirect observations; theoretical reasons for believing that an entity of such a type should exist; knowing how to produce and manipulate the entity; taking the entity to be the

[14] Hoefer (2020, p. 20): 'Briefly laying out my version of SR [Scientific Realism] will be the task in the next section. But I will have to leave a full exposition and defense of the view to future work.'

cause of a certain observable event or phenomenon (where this is not a case of observation), etc.' (Hoefer 2020, p. 32, fn. 14). Thus it seems *very* hard for Hoefer to block 'basic nuclear physics' from his examples of scientific knowledge, and yet, if we block all claims coming under the heading of 'fundamental physics (broadly construed)', it is hard *not* to block 'basic nuclear physics'—no wonder Saatsi (2020, p. 50) describes Hoefer's distinction as 'knotty'. The behaviour of atomic nuclei is 'fundamental physics', when that term is interpreted broadly. Or, if it isn't, then we end up putting a lot of pressure on a term—'fundamental physics'—that is a vague term, not built to withstand such pressure. In addition, any layperson or scientist from another field attempting to use my filter to identify future-proof science will potentially need to know a *lot* about physics to know whether a given claim is blocked or not. This goes against a core claim of this book: that non-experts should *not* think they can competently digest and assess the first-order science.

Things get worse still for Hoefer, since not only do we have trouble with the blocking of claims we don't *want* to block—we also have trouble with claims we *do* want to block. The problem now is that there are apparently claims *not* obviously coming under the heading of 'fundamental physics (broadly construed)', but which we do want to block. Perhaps the most obvious examples are those concerning *chemistry* (not physics), but where quantum effects are significant. It is well-known that quantum effects are by far most serious at tiny scales, but it is also well-known that they can emerge at the molecular level. Indeed, there is a whole field called 'molecular spectroscopy', built on the fact that molecules transition between different allowed states. What exactly is going on with such molecules depends enormously on one's interpretation of quantum mechanics. I feel sure that Hoefer would want to block certain claims about diatomic molecules, for example, since Hoefer wants to quarantine all claims that concern quantum considerations. But if such claims *do* come under the heading of 'fundamental physics', then once again we are putting huge pressure on a term that is just not designed to withstand that pressure, and will surely 'buckle under the freight' (Wilson 2006, p. 87).

Can we find an alternative way to characterise this epistemic distinction, a way that does *not* put weight on the term 'fundamental physics'? There are certainly some options in the literature. A particularly famous example is Van Fraassen's distinction concerning scientific claims about in-principle (un)observables (Van Fraassen 1980). This distinction has even been used, often enough, to define the contours of the scientific realism debate, with 'realism' standing for optimism and confidence vis-à-vis our knowledge of in-principle unobservables such as photons and electrons. But putting weight on 'observability' has been notoriously controversial in recent decades. As a way to distinguish what we can/can't have knowledge of, it seems like a poor choice given some of the examples that have emerged in the literature, of unobservables concerning which we (quite obviously, to the scientific community at least) have knowledge. Hoefer's example of 'basic nuclear physics' is one such example: the pertinent claims concern unobservables, even when we allow for a very permissive construal of 'observable'. Obviously, if 'observable' is really abused as a concept, taken very far from its literal meaning, such that it really means something else entirely, then basic nuclear physics can count as 'observable'. But the necessity to distort beyond all recognition the meaning of 'observable' just goes to show that this term is non-optimal for drawing a principled epistemic distinction.

One thing that makes the (un)observability distinction inadequate is the surprising extent of the continuity between macroscopic and microscopic behaviour. One has to get surprisingly small, it turns out, before macroscopic thinking ceases to be applicable. For example, even at the scale of molecules it turns out that common-sense assumptions about geometrical structure, about pushes and pulls, rotations, and so on, can get you quite far. It isn't illegitimate to construct a balls-and-sticks model of the double-helix DNA molecule, or the hexagonal benzene ring, for example. It was never obvious that this had to be the case: reality at that scale might have been radically distinct from reality at the scale of everyday life. But at the level of most molecules, it is *not* so radically distinct, at least in certain important respects. Whilst there may be problems conceptualising *some* features of molecules, such as the nature of molecular bonds, other features can be known, such as their basic geometry. Thus there are certain things we can know at this scale, and this goes to show that *scale* is not the right way to draw the epistemic distinction we are after. But the (un)observability distinction *does* concern scale.[15]

Another option is Stanford's distinction between scientific claims concerning domains of nature that are/are not 'far removed from ordinary human experience' (Stanford 2021). As with the other distinctions under consideration in this section, many of the most obvious examples are captured by this characterisation. The internal workings of atoms, and black holes, are 'far

[15] In fact, quantum phenomena can appear at a much larger scale, as in superfluidity. See Arndt et al. (1999), which describes the 'wave-particle duality' of C_{60} molecules; the authors note that 'C_{60} is almost a classical body, because of its many excited internal degrees of freedom and their possible couplings to the environment.' See also Shayeghi et al. (2020), who demonstrate quantum effects with even (much) larger particles. Two other nice reference points are Eibenberger et al. (2013) and Arndt and Hornberger (2014).

removed from ordinary human experience', and it would be just fine if many claims about *these* domains were blocked from passage through the future-proof filter. Stanford gives his own examples: he contrasts particle physics and cosmology (fields where there are special epistemic problems and the problem of unconceived alternatives is more serious) with ecology and geology (fields where we don't have such problems). But then, there are plenty of examples in geology where we are 'far removed from ordinary human experience' and yet we don't have special epistemic problems, including the example of the inner/outer core of the Earth discussed back in Chapter 2. Similarly when it comes to many claims coming under the heading of 'evolutionary biology', such as those explored in Chapter 4: the evolution of human beings from fish over hundreds of millions of years is most definitely 'far removed from ordinary human experience'. Whilst we *could* simply accept that lots of claims will be blocked which we don't really *want* to block, it is hardly an ideal situation. We'd be blocking claims concerning which there is an extraordinarily strong scientific consensus. The scientific community would be puzzled (to put it mildly) to see such claims blocked by the philosopher. And this is especially problematic when the distinction doesn't even enjoy a *philosophical* consensus. Put simply, there is no contest when a strong scientific consensus is pitted against a weak, or non-existent, philosophical consensus.

Other options for characterising the distinction are less well known. One option I have tried to develop concerns a 'concept application problem'. The idea is to remain sceptical whenever there is a serious question mark over the legitimacy of applying human concepts to the domain in question.[16] This way of characterising the distinction is supposed to help make sure that the problematic cases for Hoefer, Van Fraassen, and Stanford fall where they should, epistemically speaking. For example, Hoefer, Van Fraassen, and Stanford are all sceptical when it comes to claims about 'the quantum', generally speaking, including, for example, what happens during the famous two-slit experiment. This fits with the idea of a 'concept application problem': it is certainly reasonable to worry that we have a problem in attempting to apply the concepts we have—in our capacity as human beings—to quantum phenomena. That would make sense of the fact that there are so many bizarre paradoxes and radically unintuitive ideas in the field of quantum theory, including wave-particle duality, a wavefunction existing in an extremely high-dimensional space, cats in a superposition of 'dead' and 'alive', the

[16] This is closely related to Cushing's (1991) (un)picturability distinction, although his emphasis is on what we can/cannot hope to *understand*.

uncertainty principle, and the dividing universes of the many worlds interpretation of quantum mechanics.[17] One reasonable explanation of our interminable perplexity is that we just don't have adequate concepts for this context. Thus in encountering quantum mechanics we meet with a 'conceptual crisis' (Camilleri 2007, p. 198), in that our concepts are no longer fit for purpose, or as Heisenberg put it in 1958, 'these concepts do not fit nature accurately' (Heisenberg 1958, p. 55).[18]

The advantage of the 'concept application problem' (CAP) distinction is that it makes sense of the examples that frustrate Hoefer, Van Fraassen, and Stanford. When it comes to 'basic nuclear physics', for example, Hoefer wants to believe in it, but it seems to come under the heading of 'fundamental physics' (broadly construed), which means that he can't. Similarly, it seems awkward for Van Fraassen's account to *have* to be sceptical about basic nuclear physics—on the grounds that it involves unobservables—*however* strong the scientific consensus is regarding the relevant claims. Similarly for Stanford. There are claims about 'basic nuclear physics' that concern in-principle unobservables, but that we really want to allow through (given the strength of the consensus, the test of time, the abundance of 'epistemic handles', etc.). In my view, our confidence in such claims has to do with the fact that we are confident that our everyday, 'classical' concepts *can be applied* in that domain when it comes to at least *some* broad-brush claims we want to make. We can start with something simple: the claim that the atomic nucleus has a mass, in the same sense that an elephant has a mass—no CAP there. We can add the claim that it is sometimes possible to split the atomic nucleus, dividing it into pieces, as in nuclear fission of uranium-235. This 'dividing into pieces' description of nuclear fission is *not* supposed to be a metaphor, but is intended literally—in the same conceptual sense that a breakfast bowl splits into pieces when we drop it. Moreover, this simple description of nuclear fission is robust across the different interpretations of QM (cf. Fraser and Vickers, 2025). Thus, at this basic level of description, scientists would *not* tolerate talk of a CAP here, even though we are at the scale of individual atoms. To put it another way, there is no reason to put 'splits' in scare-quotes in the fission case, any more than we would put 'splits' in scare-quotes in the breakfast bowl case.

[17] As Zinkernagel (2016, p. 18) notes, 'it seems to be widely recognized that any interpretation of the theory [QM] will involve some mystery, weirdness or, at least, problematic aspects.' When we try to *force* macroscopic concepts onto quantum phenomena (e.g. the deterministic trajectories of Bohmian mechanics) something has to give somewhere.

[18] Such thoughts concerning our limits as cognitive agents have also led some to suggest that the mind–body problem will always be a paradox we are unable to solve: perhaps we just *can't* understand what's happening there (cf. O'Connor and Robb 2003).

Whilst I think the concept of 'CAP physics' is indeed helpful for catching the cases that were problematic for the other distinctions, some significant problems do arise. For one, how is 'CAP physics' to be identified? One certainly couldn't hope to apply the concept precisely when there *is* a concept application problem, since in many cases we simply *don't know* whether or not human concepts are adequate for the context in question. Thus it would need to be applied *whenever there is a serious question mark* over attempts to apply our concepts to the target object, property, or process. But historically there have been cases where the scientific community didn't think there was a CAP (e.g. the wave theory of light), but it later turned out that there was. This raises the serious possibility that cases of 'CAP physics' would be allowed through the future-proof filter, especially if (as seems sensible) we were going to trust the scientific community to tell us when there is a serious question mark over the legitimacy of applying our concepts to the target in question. We would have to face the serious threat of the filter becoming *fallible*: claims would be let through initially, only to be retracted later when scientific work revealed that we might face a concept application problem after all.

In addition, how would we even go about *asking the community* whether or not CAP applies to a given claim? Whilst there are ways of identifying a scientific consensus vis-à-vis some specific claim—and sometimes surveys are carried out with the explicit intention of measuring a scientific consensus—there aren't remotely obvious ways to identify when the scientific community would agree that CAP applies. Scientists will never have heard of 'the concept application problem'! And if non-experts, attempting to use the future-proof filter, take it upon *themselves* to identify 'CAP physics', that is highly problematic. For one thing, at the core of this book is a definite aversion to trusting one's own ability to digest and judge the first-order science. This applies just as much to identifying where a 'concept application problem' arises as it does to assessing the evidence for scientific claims.

Finally, it is hardly crystal clear what ought to count as 'not having adequate concepts'. Does a concept application problem apply to General Relativity, for example? Spacetime itself is certainly a big jump from anything we are familiar with: whilst space and time are familiar enough, the combination of the two goes radically against common-sense intuitions. And when we are invited to imagine spacetime curvature it isn't clear that we *can* imagine it. We might extend our mathematical practices from three dimensions to four dimensions (or more), but does this really help us to *imagine* four-dimensional spacetime? When we try to imagine spacetime curvature, it is tempting to imagine everyday curvature, adding 'like that, but in four dimensions'. The objection that the move from 3D to 4D is just 'a natural extension of the relevant mathematics' can be countered: after all, the introduction of imaginary numbers could be described in the same way, as could all sorts of quite bizarre mathematical constructs; professional mathematics left behind 'intuition' a very long time ago. But on the other hand, Cushing (1991, p. 343) states, 'In a sense, Einstein's general theory of relativity provided an understandable (picturable) causal explanation in terms of a curved space-time.' No doubt *some* physicists who have spent many years working with General Relativity would insist that one can 'get used to it', *acquiring* concepts such as *spacetime curvature* that the vast majority of human beings never acquire. Whether CAP applies or not is unclear, and we might want to allow for borderline cases where our concepts apply up to a point, but we are being asked to stretch them. But the future-proof filter needs to make an absolute decision: will such claims be blocked, or not? Stating that claims are blocked whenever it is *unclear* whether or not CAP applies, or whenever our concepts are being *stretched*, will just move the problem along, without solving it.

Thus, once again, the distinction seems either inadequate, or fragile. We could now search around for other ways of characterising the distinction, other concepts to put the epistemic weight on. We might put weight on 'non-classical physics', for example. But on close inspection this crumbles, too. The term 'classical physics' is both plural and fuzzy in its application. It is plural in the sense that it is used in different ways by different scholars, and it is fuzzy in the sense that it is often used in a vague way, without anything like a clear definition. Gooday and Mitchell (2013, p. 727) identify four different ways 'classical physics' is typically used. They conclude that, 'as a descriptive tool, "classical" ought to be abandoned altogether. Even if carefully defined, the term is so ingrained amongst historians and philosophers of physics that it is liable to lead to confusion.' (p. 729). I am certainly sympathetic (Vickers 2013b, 2014; Taylor and Vickers 2017) to this kind of terminological eliminativism.

We might keep searching. But alternatively we might wonder whether *any* way of making a gross generalisation concerning the *kinds* of claims that should always be doubted is going to come up against the same three objections:

(i) The key term used to define the distinction ('fundamental physics', 'in-principle unobservable', 'concept application problem', etc.) is always going to be too plural, vague, or contested to withstand the epistemic weight placed upon it.

(ii) The distinction will always be controversial, inviting counterexamples of one or both kinds: (a) claims allowed that ought to be disallowed, and (b) claims disallowed that should be allowed.
(iii) Deciding what comes under the heading 'fundamental physics', 'in-principle unobservable', 'concept application problem', and so on, is always going to require specialist *first-order* scientific knowledge. But the future-proof filter is supposed to focus on *second-order* knowledge, such that users can identify future-proof science without needing to understand the first-order science in question.

Thus we might wonder whether we can somehow manage without introducing a crude distinction between claims of a certain kind, or coming from a certain field, that should always be doubted, and other claims.

The idea behind introducing a distinction was this: it is reasonable to be sceptical of all scientific claims on one side of the distinction, given their fundamental character (even without starting to explore the evidence for/against them). Why is it reasonable? Three reasons are:

(1) History shows that ideas coming out of certain corners of 'fundamental physics' (broadly construed) are more vulnerable to periodic scientific revolutions than ideas coming out of other fields.
(2) The problem of unconceived alternatives is more serious in fundamental physics (broadly construed) than in other fields.
(3) In certain corners of fundamental physics we ought to seriously doubt that we even *have the concepts* to describe what is going on.

But why not suppose that a sufficient number of members of the relevant scientific community are perfectly aware of such reasons for caution? Meaning that, where appropriate, they will remain cautious, and the 95 per cent benchmark won't be reached? If that's true, then the criteria for future-proof science introduced in Chapter 5 will already block all the claims we want to block.

4. Rejecting Calls for a Principled Epistemic Distinction

Aren't modern-day theoretical physicists well aware of how cautious they need to be vis-à-vis claims about 'known facts' in their field? Don't they know the history of their field well enough to know that, in non-classical theoretical physics, things are often not what they seem, and (nearly) all bets are off when it comes to possible surprises in their field? If so, then solid 95 per cent consensuses coming out of fundamental physics (broadly construed) are going to be rare.

Consider quantum theories, including QM and QED. It is often said that such theories are *incredibly* well confirmed, and should thus be right at the *top* of the list of the 'scientific realist' who is concerned to identify scientific truths. We already saw how Worrall (2007) draws our attention to the amazing predictions of QED, as highly pertinent to the realism debate. Callender (2020) describes the scientific realist as 'someone who believes that mature successful theories are well confirmed and approximately true' (p. 72). And we certainly find 'maturity' and 'success' within theoretical physics, including QM. Since the different interpretations of QM actually are different theories (Callender 2020, p. 59; also Hoefer 2020, p. 27) the realist is in trouble, since she apparently must believe in the (approximate) truth of at least three thoroughly inconsistent theories (Bohm, Everett, and Collapse), thus contradicting herself. Even the qualifier 'approximate' can't help her, since the theories differ too greatly. Callender writes,

> One can't defend a belief in collapses if one at the same time admits that the evidence equally well supports a theory without them. Are we really in this situation? Prima facie, yes. [...] [T]hese three programs are neither philosophers' toys nor notational variants (on any remotely reasonable semantics) and are clearly 'scientific' in letter and spirit. Quantum underdetermination is the real deal. (2020, p. 61)

He concludes: 'In sum, we have serious scientific underdetermination. The nightmare of scientific realists is real' (p. 75).

But note how easy it is to sidestep this problem if one is not engaged in the realism debate, as I am not (recall Chapter 2). I'm instead attempting to identify (some, not all) future-proof science, and I have argued that the best way to do this is to identify a solid scientific consensus, meeting the criteria put forward in Section 3 of Chapter 5. This involves a commitment to the relevant scientific claims being considered 'facts' by that community (not merely 'promising theories', say). And there is nothing remotely approaching this when it comes to the interpretations of QM Callender discusses.[19]

[19] See Fraser and Vickers (2025) for further discussion.

How *might* quantum mechanics pose a problem for my account? In an alternative counterfactual history, mightn't the community have fixated on one interpretation to such an extent that other possible interpretations never got off the ground, or were considered definitely 'fringe'? Indulging in some counterfactual history, Saatsi (2020) writes as follows:

> Assume, for the sake of the argument, that in the scenario envisaged here [an alternative history where the de Broglie–Bohm theory is the only interpretation of QM 'on the table'] most physicists are Bohmians about quantum mechanics, and that relativistic and field-theoretic extensions of Bohmian mechanics are the hottest area of research around. Such broad allegiance to Bohmianism should not convince the realist, since the commitments of truth–content realism are not read off from scientists' beliefs. (p. 62, fn. 19)

And yet, in this book I *am* (pretty much) suggesting that examples of future-proof science should be 'read off' from scientists beliefs, as opposed to attempting any kind of assessment of the first-order evidence (recall Chapter 4); not individual scientists of course, but when there is a certain kind of very strong scientific consensus, as defined by the two criteria in Section 3 of Chapter 5. Saatsi might worry about this as follows: if the history of QM *had* proceeded with Bohmian mechanics taking precedent to such an extent that other interpretations weren't even identified, then there would have been a strong consensus vis-à-vis certain highly contentious claims, such as that electrons are particles that always follow deterministic trajectories.

However, Saatsi here slides from a scientist identifying as a Bohmian, to a scientist *believing the Bohmian story*. These are very different things. I think it is reasonably plausible that—had things gone differently—most physicists might have identified as Bohmians. I think it is highly *im*plausible that—had things gone differently—the vast majority of physicists might have asserted that the Bohmian picture of the quantum world is a 'known fact', in the way scientists make assertions regarding the 30 examples given in Chapter 1. Since only a tiny percentage (perhaps even 0%) of modern-day, actual Bohmians would say that the Bohmian story is 'known fact', it is unreasonable to suppose that 95 per cent of *all* relevant physicists might ever have said this, even if Bohmian mechanics were the only known, or only 'serious', interpretation.

This idea gets support from the fact that the quantum revolution—famously—brought scientists to agonise over deep questions concerning the limits of human knowledge. It is surely uncontroversial to claim that lessons were learned during this process, lessons that would forever influence community confidence in claims coming under the heading of 'fundamental physics'. In particular, the idea that the concepts we have—in our capacity as human beings—are not suited to describing certain domains became a much more serious idea than previously. As Heisenberg (influenced by Bohr) put it in 1958,

> The concepts of classical physics are just a refinement of the concepts of daily life and are an essential part of the language which forms the basis of all natural science. [. . .] There is no use discussing what could be done if we were other beings than we are. (1958, p. 55)

The latter sentence is crucial: for Heisenberg and Bohr, to think humanly just is to employ classical concepts (cf. Cushing 1991, p. 355), concepts that are *not* suited to quantum physics. Consider also MacKinnon (1982, p. 271) discussing Bohr's position:

> These terms ['position' and 'trajectory'] already have a determined meaning prior to and independent of any particular atomic experiments. The basic meaning of these terms is determined by ordinary language usage proper to classical physics. The critical problem requiring analysis accordingly is how such already meaningful classical concepts function in quantum contexts.

For those in the Copenhagen school, especially Bohr and Heisenberg, 'classical' referred to pre-quantum, and *also* to the concepts we can and do actually employ, as human beings. Thus quantum theory presented a fundamental barrier to knowledge. As Landsman (2006) notes, 'Bohr's doctrine of classical concepts implies that no direct access to the quantum world is possible, leaving its essence unknowable' (p. 218).[20] This philosophy was popular enough, at the time, for us to confidently assert that at least 5 per cent of the relevant community would have dissented from any claim to a 'future-proof fact', even in a (realistic) counterfactual history.

Thus interpretations of QM are no problem for my account, whether we consider the actual history of science, or (realistic) counterfactual histories. But consider other examples, where a solid consensus is perhaps more plausible. Perhaps the most obvious examples are those from the 19th century coming under the broad heading of 'classical physics', including the wave theory of light, classical mechanics, and corresponding background assumptions such as that

[20] Some scholars would disagree with this interpretation of Bohr's philosophy of QM, but this is at least a reasonable interpretation of at least many of the things Bohr said. If preferred, it could be called a 'Bohrian' doctrine.

space is Euclidean and space and time are independent of one another. Isn't it reasonable to suppose that there was a solid consensus regarding such things, in the relevant community, in the 19th century?

One option here would be to admit that the future-proof filter would *not* have worked back in the 19th century—it would have allowed through claims that were *not* future-proof—but to insist that the filter is reliable *now*. But this isn't necessary, and I want to hold on to one of the key claims at the heart of this book, namely,

> *No scientific claim has ever—in the entire history of science—met my criteria for future-proof science and then subsequently been overturned, despite enormous opportunity for that to happen (were it ever going to happen).*

How is this assertible in the face of 19th-century classical physics? To be blunt, when we think about 19th-century classical physics, relevant claims (e.g. concerning the nature of light) did *not* meet my criteria for future-proof science for at least one reason: the community wasn't sufficiently diverse, for example because 99 per cent (give or take) of the relevant scientists were male. This may seem like a 'cheap' way to wriggle out of possible counterexamples, and a critic could speculate that even a gender-balanced physics community in the 19th century would have been confident that classical physics was future-proof. But this is pure speculation, and is based on the controversial assumption that scientific community diversity *of a kind that includes some measure of gender diversity* isn't a particularly important consideration when it comes to epistemology. In any case, I don't need to fight this fight: the fact remains that claims from 19th-century physics don't make it through the future-proof filter. And, of course, the 'diversity' criterion is independently motivated—see, for example, Longino (1990, p. 76; 2002, p. 131f.) and Oreskes (2019, p. 143f.)—not included merely to rule out counterexamples from 19th-century physics. (For more on the diversity requirement for future-proof science, see Chapter 9, Section 2.2.)

Thus we reach the thought that we can keep the future-proof filter simple, adopting the criteria of Chapter 5 for the *whole* of science, without worrying about making adjustments for certain scientific fields (e.g. 'fundamental physics'), or contexts (e.g. where quantum effects are significant), or claims of a certain kind. I agree that there are special considerations in some contexts, but there is good reason to believe that the relevant scientific communities are sufficiently aware of these 'special considerations', such that we can trust their judgement in these contexts just as much as in any other. It is worth stressing again that 95 per cent consensus is a very high bar, and when I say 'trust their judgement' it doesn't mean following the scientific community when there is merely a 'strong' consensus of, for example, 80 per cent. It might be right that a large number of physicists are often *too confident*, or have forgotten the lessons from history, or never learned them in the first place. But to fall short of the 95 per cent threshold it is only required that 5 per cent of the community have doubts, and that modest degree of doubt is very easy to find in the context of theoretical physics. In addition, '[b]y nature, scientists are a conservative and cautious bunch' (Boudry 2020)—recall the discussion in Section 3 of Chapter 5.

As well as keeping the filter relatively simple, the 95 per cent approach applied to 'fundamental physics' digs Hoefer (2020) out of the hole he dug for himself when he advocated quarantining claims within 'fundamental physics' (broadly construed), but also wanted to claim knowledge about 'basic nuclear physics', as well as stating that 'the properties and behaviors of the atoms and molecules are understood in great detail' (Hoefer 2020, p. 22f.). Without conducting a case study here, I submit that basic nuclear physics enjoys at least 95 per cent consensus within the relevant scientific community. Relevant claims include: 'In nuclear fission of uranium-235 the uranium nucleus divides into pieces, releasing energy in the process.' Such a claim can be found in all of the relevant textbooks, a good sign that there is a solid scientific consensus amounting to at least 95 per cent.

None of this is to deny that 'fundamental physics' is an area where special caution is to be recommended, where there is a 'special vulnerability to underdetermination' (Hoefer 2020, p. 21). Fundamental physics *is* a special case. But crucially, the special caution doesn't have to be *built into* the future-proof filter if we believe that special caution is already applied in the course of the relevant scientific community reaching a 95 per cent consensus, and being willing to use the word 'fact'. In other words, we might well suppose that the relevant scientific community is sufficiently aware of the lessons from history, and the worry that human concepts may sometimes be inapplicable within a certain domain of investigation, such that a 95 per cent consensus just is *already* a higher bar within the relevant context of scientific endeavour, even without us manually raising the bar to 97 per cent or 98 per cent (say) in this context.

5. Interpreting Claims from Fundamental Physics

Like me, Saatsi (2020) rejects a principled epistemic distinction in his discussion of quantum spin. He writes:

> Spin, a theoretical concept at the heart of quantum theory, is scientifically as firmly established as any. Spin underlies numerous explanations and predictions in physics and chemistry, as well as a rapidly growing number of technological feats. Its central and multifaceted theoretical role strongly motivates a scientific realist attitude: we should be "realists about spin"—as much as we are realists about anything in physics. (p. 47)

And he goes on to describe the numerous ways in which 'spin' has been an incredibly successful theoretical concept, concerning each of (i) its theoretical virtues, (ii) its empirical virtues, and (iii) its employment in the development of new technologies, such as those coming under the heading of 'spintronics'. Of course, I would prefer the second-order-evidence route, focusing on the measure of consensus of the relevant scientific community. But Saatsi and I do at least agree that we should *believe in spin*.

However, what does it mean to say, for example, that electrons have spin? Our layperson, making use of the future-proof filter, may be able to establish that this claim gets through the filter, and is thus future-proof. But, as things stand, that layperson may have no idea what 'spin' means in this context, or may even assume that the everyday concept of 'spin' is applicable, when every relevant scientist knows that it is not. Thus, whilst the layperson gets the *statement* right—'electrons have spin'—she potentially gets the *proposition* very wrong: she thinks the statement means something it does *not* mean when she imagines a tiny spinning ball.

There are numerous similar examples within fundamental physics. Terminological convenience means that scientists often describe things in ways that are not meant literally. Physicists still talk about electron 'orbits', for example, whilst completely rejecting the possibility that electrons literally orbit the atomic nucleus (cf. Callender 2020, p. 69). Three specific examples are Biedenharn (1983, p. 21), Letokhov and Johansson (2009, p. 37), and DeMille (2015, p. 35); the latter writes,

> Even diatomic molecules can display a similar effect, if they have an electron orbiting the internuclear axis with a nonvanishing angular-momentum component Ω along the axis. In that case, a mirror-image state with the electron orbiting in the opposite direction would nominally have the identical energy. (p. 35)

But are the electrons actually 'orbiting'? Definitely not; all physicists are agreed. Biedenharn (1983) at least clearly flags up that he is not using the term 'circular orbits' literally—he uses the term to refer to cases of 'nodeless radial probability density' (p. 21).

Thus we should be very cautious when we interpret the language of physicists, and this includes any claims coming out of fundamental physics that happen to pass through the future-proof filter. My proposal for the layperson who would wish to make use of the filter to identify future-proof science is this: she should consult relevant textbooks on *how to interpret* claims passing through the filter that come out of fundamental physics. And how should she identify which claims *do* come out of fundamental physics? Crucially, we shouldn't here invite back into the picture the difficulty for the layperson of hoping to competently distinguish between physics and fundamental physics. Instead, the layperson should consider the names of the journals, and textbooks, where the claim in question is predominantly to be found. If these names include terms such as 'fundamental physics', 'quantum physics', 'non-classical physics', 'theoretical physics', 'quantum chemistry', 'foundational physics', or similar, then the layperson is advised to consult expert help to interpret the meaning of the claim in question. Sometimes the claim won't need any specialist interpretation, as when nuclear fission is described as splitting an atomic nucleus into parts. Asking for advice on interpretation in such cases won't harm, however. Thus the layperson might as well consult relevant textbooks or ask for advice interpreting any claim where there is even the slightest chance of significant misinterpretation. The layperson should also be prepared to accept that she might be unable to understand the interpretation she is offered (e.g. the meaning of 'spin'). It is by no means guaranteed that a non-expert will have at hand the concepts needed to understand a claim coming out of fundamental physics.

A crucial distinction between claims about 'orbits' and claims about 'spin' now needs to be factored in. Claims about 'orbits' will *not* make their way through the future-proof filter, and so the question of how to interpret such claims won't arise. This is because physicists' tolerance for making claims that are misleading when taken at face value can only be pushed so far. If that tolerance was truly unlimited, then claims about electron orbits would feature throughout the literature, and caveats about the *real* meaning of such talk (e.g. 'nodeless radial probability density') wouldn't even be deemed necessary. And when prompted, relevant experts would say that it is a scientific fact that electrons orbit the nucleus of an atom, whilst bearing in mind what the word 'orbit' really means in this context. One does *not* find this widely in the literature, and far less than 95 per cent of the relevant community would describe 'electrons orbit the nucleus' as a scientific fact. Indeed, perhaps

95 per cent of the community would say it is a scientific fact that electrons do *not* orbit the nucleus. The word 'orbit' has largely retained its classical meaning, without taking on board a new, quantum meaning in the way the word 'spin' has.

When it comes to 'spin', the idea that quantum spin is a property of electrons, photons, quarks, neutrinos, and various other particles is absolutely established, and found in all of the relevant textbooks that discuss the Standard Model of particle physics. But it may be objected, what if the Bohmian interpretation of QM is true? On this interpretation spin is *not* a fundamental property of electrons, but instead reduces to other properties and relations (see e.g. Daumer et al. 1997). Since this is an interpretation that is taken quite seriously by the relevant community, shouldn't we be concerned about letting claims such as 'electrons have spin ½' through the future-proof filter? But in fact this isn't a problem. Even if Bohmian mechanics is the true story of QM, it will still be correct to say that electrons have spin ½. As discussed in Egg (2021, Section 4), it remains the case that electrons have spin ½, even if, on the Bohmian account, what makes this statement true differs starkly from what makes it true on alternative interpretations of QM. Bohmian mechanics certainly involves a spin operator, and Bohmians do still talk about 'spin' quite freely. They just have a different account when it comes to the more fundamental story (cf. Saatsi 2020).

Another interesting case illustrating the challenges of interpretation concerns claims about diatomic molecules. Now, when we hear the word 'molecule' we may imagine that we are firmly in the context of *chemistry*, and in the vast majority of chemistry the paradoxes of quantum physics are not relevant. But in some cases quantum effects *are* significant. Diatomic molecules are a case in point (Figure 6.3).

Miyake and Smith (2021) challenge Van Fraassen, arguing that diatomic molecules are unobservables we certainly have knowledge of, where the justification comes via the quantity and quality of the evidence over many decades. Miyake and Smith emphasise: 'There is now an online database of papers on diatomic molecules... it contains 30,000 references for the period 1974–2000.' (p. 164). Given the exponential growth of science, one can add at least 30,000 additional relevant publications for the period 2000–22, bringing us to 60,000+ overall. As the authors note: 'one virtue of focusing on diatomic molecules is the amount of evidence that has been developed over the course of the last 90 years of research on them—at least comparable to, if not substantially exceeding, the scope of the evidence developed over the last three centuries in celestial mechanics' (p. 178). Thus we apparently have a

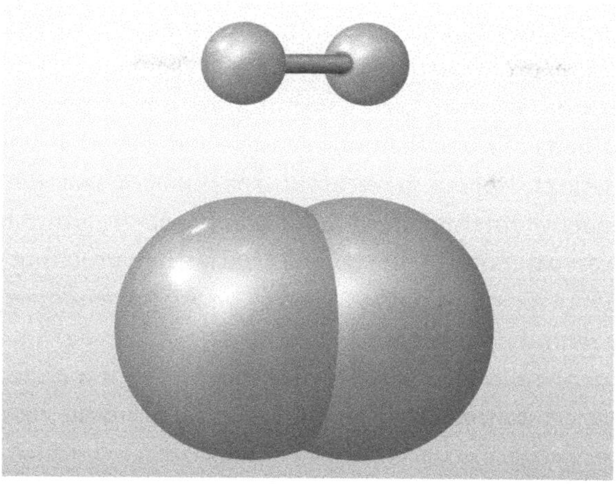

Figure 6.3 Diatomic hydrogen

situation where claims about diatomic molecules are totally solid, beyond all reasonable doubt, and yet Van Fraassen would counsel caution, purely on the grounds that they are unobservables. Miyake and Smith would surely reply that the fact that they are unobservables is irrelevant, since the last 90 years of research shows just how scientifically accessible they are, despite being unobservable. Hoefer would say that when it comes to diatomic molecules, there is an abundance of 'epistemic handles', such that we crossed the threshold for knowledge many years ago.

Various claims about diatomic molecules would get through the future-proof filter. There are 'scientific facts' concerning the rotation of such molecules, the average distance between the two nuclei, and the energy required to pull the two nuclei apart (just to give three examples). For example, it is a scientific fact that the average 'equilibrium' distance between the two nuclei in normal diatomic hydrogen is $0.7416x10^{-8}$cm (with small uncertainty). But at the same time caution is required. It is well-known that diatomic molecules make 'quantum jumps' between different allowed states; the fact that they do this is the very basis of molecular spectroscopy. What exactly is happening when such a molecule transitions between two states? This is far from clear, and different interpretations of quantum theory provide very different answers. In the Bohmian interpretation there actually isn't a 'jump' at all; particles follow deterministic trajectories. Whatever is really happening, we are well within the realm of quantum effects here. One may suggest that when the molecules are *in a definite state*, in between transitions, their nature is knowable: for example, they rotate in the same sense that a thrown stick rotates. But this too is more problematic than it initially sounds. On some (popular) interpretations of QM there are states of a diatomic molecule which are in fact superpositions of more than one possible state. In what sense, now, is the molecule rotating? Does it have a superposition of two different rates of rotation, somehow rotating slowly and quickly at the same time? Here we lose our conceptual footing—the comparison with the rotating stick suddenly seems more like a loose analogy, and less like a way to genuinely visualise what is happening with the molecule.

What about the distance between the two nuclei of a diatomic molecule? This is constantly shifting; Miyake and Smith (2021) write that,

> [T]he molecular constants are a way of capturing a situation that is enormously more complicated, with time-averaging smoothing or glossing over fluctuations of much shorter periods. (p. 172)

But perhaps we can specify the *average* distance, just as we understand what it means to say that the average distance between the Earth and Sun is 149.6 million km? Indeed, Miyake and Smith encourage us to accept that it is a scientific fact that the time-averaged 'equilibrium' distance between the two nuclei in normal diatomic hydrogen is $0.7416 x 10^{-8}$cm.

But, once again, what of superposition states? A given diatomic molecule can—according to some, popular, interpretations—be in a superposition of two possible states. This allows for the distance between the two nuclei to be a superposition of two different distances. Here we are clearly being asked to stretch our everyday concepts considerably, and we may even feel the conceptual floor drop out from beneath us. We can't visualise such a superposition state and so it is reasonable to say that we can't actually understand what is happening (Cushing 1991). In which case, when we confidently assert that it is a scientific fact that the average distance between the two nuclei in normal diatomic hydrogen is $0.7416 x 10^{-8}$cm, we are suggesting a greater degree of understanding than is actually justified. We are speaking as if everyday, macroscopic concepts can be straight-forwardly applied, when it is quite possible they can't. Miyake and Smith (2021) make a comparison with time-averaging in celestial mechanics, but whereas in celestial mechanics we understand what is happening instantaneously, at any given time, this can't be confidently asserted of diatomic hydrogen.

None of this affects the fact that such claims—about diatomic molecules' rotations, energies, and so on—make it through the future-proof filter. They are established scientific facts, and that huge wealth of literature referenced by Miyake and Smith (2021) testifies to that. The caution comes in at the interpretation stage, *after* the claim has passed through the future-proof filter. The layperson needs to ask the expert how these claims should be interpreted. And the expert will say something about how the average 'equilibrium' distance between the two nuclei in normal diatomic hydrogen is 'effective', and 'glosses over a situation that is enormously more complicated' using the tool of time-averaging. Similarly when it comes to the rate of rotation.

This isn't to say that *all* claims about diatomic molecules need specialist interpretation. For example, we do know that there are two nuclei, and we know the energy required to pull those nuclei apart ('dissociation energy'), for all of the different diatomic elements. For example, the energy required to break the bond of normal diatomic hydrogen, in normal conditions, is 435.7799kJ/mol, with very small uncertainty (Luo 2007, p. 3). And when we talk about the energy required to pull the nuclei apart, we are talking about the same *kind* of thing we are already familiar with (e.g. when we pull apart two strong magnets that are stuck together). Thus, there isn't a concept application problem. The layperson doesn't judge this for herself, however; she must ask the expert. How does she *know* that she needs to ask the expert, in this case? For one thing, when we consider diatomic molecules, we are very close to the level of the individual atom, which is an exemplar of 'fundamental physics' (broadly construed). For another, claims about diatomic molecules are often found in the journals *International Journal of Quantum Chemistry* and *International Journal of Molecular and Theoretical Physics*, and both journal names are red flags for obvious reasons.

I certainly don't hope to here solve the problem of how to interpret claims coming out of 'fundamental physics'. That would be a very different book, and what I say in this section is tentative. My proposal is to trust claims—even those coming out of 'fundamental physics'—where the future-proof criteria are met, but to be cautious about how to interpret those claims. I've suggested that the non-expert might learn how best to interpret them by consulting relevant textbooks concerning key terms such as 'spin'. If there is no consensus on how to interpret the key terms, then I submit that the claim wouldn't get through the filter in the first place. This is because at least 5 per cent of the relevant community would *not* be willing to describe the claim as an indisputable 'fact', since they would insist that it depends on what the statement is taken to mean, and—being aware of the controversy—they would note that different options are available. Recall that I am using a rather strong sense of 'fact', where those agreeing to use the word 'fact' must be willing to agree that debate on the matter should be shut down, with dissenting articles declined by the relevant journals (just as we find with articles opposing climate change and evolution).

6. Conclusion

My approach to 'fundamental physics' (broadly construed) can be summarised as follows.

Fundamental physics is a special case for two reasons: (i) revolutions in scientific thinking are more likely in this context than in other scientific fields, so we need to be comparatively more cautious, and (ii) claims coming out of fundamental physics often require especially careful interpretation, since they often don't wear their meaning on their face. I have argued that neither (i) nor (ii) warrants introducing a special 'block' as part of the future-proof filter (à la Hoefer 2020), refusing passage to all claims coming out of fundamental physics. Whilst we do need to be more cautious, the relevant community is aware of that, meaning that 95 per cent consensus in this context is *already* a higher bar than in other contexts. Since the community is more cautious, there is no need to raise the bar higher than 95 per cent, or introduce special conditions.

When it comes to (ii), the care with interpretation can happen *after* a claim has passed through the future-proof filter, since for present purposes (in this book) we are ignoring claims that do not pass through. Even taking care with interpretation, the non-expert may find it difficult or impossible to understand what is meant by a given claim, although they might make headway by consulting accepted textbook definitions of key terms. Textbook definitions of key terms (e.g. 'spin') can be trusted to the extent that the definitions are consistent across textbooks, because textbook writers work hard to make sure their definitions of key terms are uncontentious (see Chapter 9, Section 2.4, for further discussion of textbook definitions). At the same time, this book isn't particularly concerned with the extent to which non-experts can understand mind-bending concepts in physics. It's undeniable that sometimes laypersons simply won't be able to understand claims coming out of fundamental physics. Some concepts in fundamental physics are *particularly* difficult to grasp and require years of study to properly acquire, including *spacetime curvature* and *quantum entanglement*. Thus, on my account, one can know that a claim is future-proof, without knowing what the claim means.

Despite important open questions, I leave the discussion of fundamental physics behind for now. That context of scientific theorising is certainly not the primary focus of this book, since the vast majority of future-proof scientific claims will come from other scientific fields. And there remain important questions to ask vis-à-vis future-proof science having nothing to do with fundamental physics, especially as concerns contemporary cases the reader may be curious about. One such case is the asteroid impact account of the extinction of the dinosaurs. In this case, worries about fundamental physics are unimportant, and other issues come to the fore. In particular it must be asked: How can the individual *actually ascertain* whether there is the kind of solid scientific consensus that would make for future-proof knowledge? A worry arises that it isn't possible, in practice, for the individual to know whether there is the right kind of scientific consensus or not. In addition, the individual—in trying to cut corners in order to reach a decision—may be led astray by the information that is most directly accessible. It is to these issues we now turn.

7

Do We Know How the Dinosaurs Died?

1. Introduction

This book puts *scientific consensus* at the heart of identifying future-proof science. Suppose it is correct. Suppose the criteria put forward in Section 3 of Chapter 5—a 95 per cent consensus of opinion amongst relevant experts embedded within a diverse, international scientific community—are indeed sufficient for a scientific claim to be future-proof, and that there are many examples of scientific claims meeting those criteria. It doesn't follow that the curious individual (whether layperson *or* expert) is able, in practice, to ascertain whether there is a sufficiently strong scientific consensus in a given case. We must ask: How exactly is the individual supposed to find out whether or not 95 per cent of relevant experts would call the claim in question a fact?

The question of what caused the End-Cretaceous mass extinction—approximately 66mya—makes for an interesting and instructive contemporary case. A very significant causal factor for the extinction event is demanded, since it is known that somewhere between 40 per cent and 76 per cent of all species, and approximately 14 per cent of all families, including virtually all of the dinosaurs, went extinct at that time (Jones 2006, p. 258). Losing all of the non-avian dinosaurs is a really big deal: they had been around for approximately 170 million years, which is a gargantuan amount of time. Many of them would have had an impressive capacity to adapt to new climates and environments. This impressive capacity is demanded by the fact that, in those previous 170 million years, dinosaurs would have been tested by environmental disruptions on many occasions. Flexible species would have been rewarded for their flexibility, with inflexible species dying out. Specifically, as Brusatte et al. (2015, p. 629) note,

> Dinosaurs survived a mass extinction at the end of the Triassic that had little clear impact on their diversity, as well as a poorly understood, but possibly important, extinction event at the end of the Jurassic.

Identifying Future-Proof Science. Peter Vickers, Oxford University Press. © Peter Vickers 2023.
DOI: 10.1093/oso/9780192862730.003.0007

Thus the dinosaurs living 66mya were the descendants of species that had prospered in the past precisely because of their ability to adapt in the face of adversity. But whatever happened 66mya, the non-avian dinosaurs could not adapt. They *all* died.

As mentioned briefly back in Chapter 2, in the 1970s several possible explanations were more-or-less on the table, including:

- Climate change
- Supervolcanoes
- Asteroid impact (the 'Alvarez hypothesis', after Luis and Walter Alvarez)

After 1980—specifically the publication of Alvarez et al. (1980)—it became common to find bold statements that an asteroid was responsible, without much (or any) consideration for the other options. As Courtillot (1999, p. 138) puts it, 'Not only editors of magazines for a broad public, but even those of some major scientific journals like *Science*, quickly opted in favor of the impact theory.' More specifically the suggestion was this: an asteroid was causally responsible for the K-Pg extinction; without the asteroid there wouldn't have been *anything like* the actual mass extinction that took place. Perhaps there would have been *some kind* of (mass) extinction event, due to other causes; as I will soon discuss, 'Deccan Traps' volcanism was causing significant disruption 66mya, and this volcanism might have caused a lot of trouble for species *on its own*, even without the asteroid. But, according to Alvarez supporters, that would have been a very different kind of extinction event (probably much slower and much less severe). At the heart of the Alvarez hypothesis is a claim that an asteroid impact was a *very* significant causal factor in the demise of the dinosaurs. Courtillot (1999, p. 142) writes, 'while the Deccan Traps were already erupting and the biosphere was sorely put to the test, an impact occurred at the same time [...]. This is what seems most likely today.' What the advocates of the Alvarez hypothesis stick by, however, is a claim that without the asteroid things would have been *vastly* different.

Today, just as in the 1980s, it is easy to find bold statements that an asteroid was responsible. Parents sharing the wonderful *Encyclopedia of Animals* (Howard and Vogel 2019) with their children will read on page 48 that, 'fish have survived many mass extinction events including the meteorite that killed the dinosaurs.' Looking online, many articles aimed at a 'popular' audience state the Alvarez hypothesis as a fact. For example, recent online articles include, 'Dino-Killing Asteroid Hit Just the Right Spot to Trigger Extinction'

(2017),[1] 'The Asteroid Impact That Doomed the Dinosaurs Also Spawned a Mega-Tsunami' (2019),[2] '66-million-year-old deathbed linked to dinosaur-killing meteor' (2019),[3] and 'Scientists find origin of asteroid that killed the dinosaurs – and another one could be on its way' (2021).[4] In addition, a simple internet search leads one quickly to a Wikipedia page entitled 'Alvarez hypothesis',[5] which includes the following:

> In March 2010, an international panel of scientists endorsed the asteroid hypothesis, specifically the Chicxulub impact, as being the cause of the extinction. A team of 41 scientists reviewed 20 years of scientific literature and in so doing also ruled out other theories such as massive volcanism.

...where 'Chicxulub impact' refers to whatever caused the 150km-wide impact crater discovered in 1990, buried beneath the Yucatán Peninsula in Mexico. Whilst Wikipedia is not always a good source of information, a reference is included to a bona fide scientific publication (Schulte et al. 2010), in the journal *Science*, one of the most reputable scientific journals in the world. Turning to Schulte et al. (2010) it is easy to confirm that there are 41 authors, and that the opinion voiced by the article is truly international, since the authors come from a wide range of countries. The reader may next look to see just how sure the authors are. And already in the abstract one finds simply, '[T]he Chicxulub impact triggered the mass extinction.' There is no 'probably', and apparently no doubt.

Initially this seemed to me quite convincing. To confirm it I asked a couple of relevant scientists I happened to know. One suggested, 'There have been 2 or 3 papers in the last few years that I think have really ended this argument [in favour of Alvarez]', and another directed me to that 2019 online article '66-million-year-old deathbed linked to dinosaur-killing meteor', adding, 'So this more or less clinches it for the meteorite.' The research linked to the 2019 article concerns a site dubbed 'Tanis' in modern-day North Dakota, where very strong evidence has been uncovered of the devastating effects of an

[1] See https://www.nationalgeographic.com/science/article/dinosaurs-extinction-asteroid-chicxulub-soot-earth-science (last accessed March 2022).

[2] See https://www.space.com/dinosaur-asteroid-chicxulub-caused-massive-tsunami.html#:~:text=The%20dinosaur%2Dkilling%20asteroid%20that,the%20Yucatan%20Peninsula%20in%20Mexico (last accessed March 2022).

[3] See https://news.berkeley.edu/2019/03/29/66-million-year-old-deathbed-linked-to-dinosaur-killing-meteor/ (last accessed March 2022).

[4] See https://www.independent.co.uk/space/asteroid-dinosaurs-earth-crater-crash-kill-b1899942.html (last accessed March 2022).

[5] See https://en.wikipedia.org/wiki/Alvarez_hypothesis (last accessed February 2022).

asteroid impact. It certainly seems undeniable, given this new evidence, that a major impact event led immediately to the demise of the dinosaurs at that site. The details are startling: one can effectively 'see' how particular organisms at that site died, on the very day the asteroid hit. For example, fossils of fish have been uncovered with their gills clogged with 'tektites'—pieces of glass created and thrown up by the asteroid impact, which then rained down on the Tanis site (Figure 7.1: image A, top left, shows a recovered sturgeon skull fossil; image B, top right, is a close-up of part of that skull, showing the tektites in the gills; images C and D, bottom, show microtektites in the gills of another fossilised fish recovered from the site).

But despite all this—the online 'popular' articles, the Wikipedia page, the 2010 review article in *Science*, and the opinions of those two relevant scientists I happened to know—we are some distance from establishing a 95 per cent solid international consensus. It may be *true* that satisfaction of my two criteria is sufficient for future-proof science, but it is important to supplement that with a good procedure for actually *identifying* when such a consensus exists, and when it doesn't.

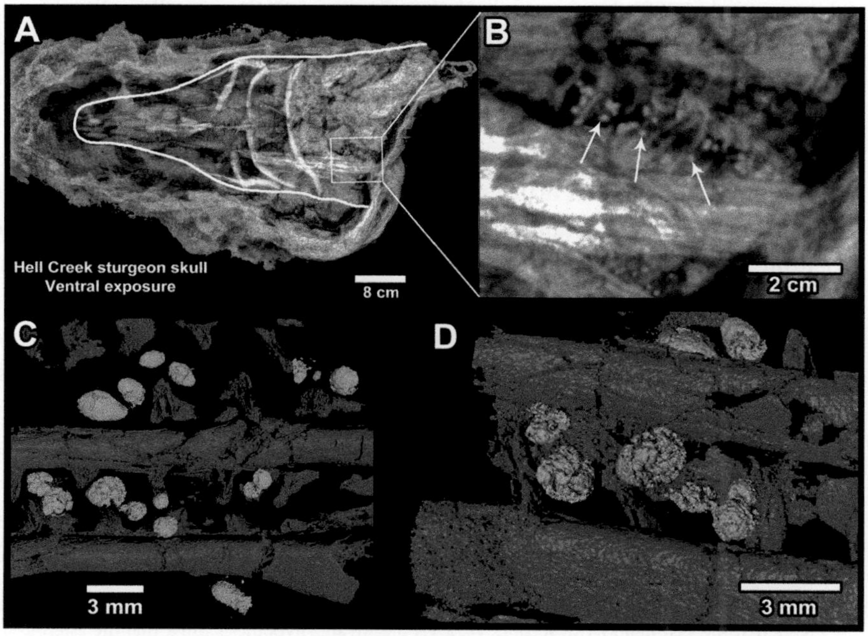

Figure 7.1 Fossilised fish with their gills clogged with tektites at a site coinciding with the K-Pg boundary

2. Assessing the Opposition—First Pass

Let's start with the scientists I happened to have access to, and that off-the-cuff remark, 'this more or less clinches it for the meteorite'. I think we can all be tempted by the promise of silver bullet evidence sometimes. We saw earlier that even Einstein made a knee-jerk response to Bohr's predictive success vis-à-vis the spectral lines of ionised helium in 1913. But such responses should not be taken too seriously, and it isn't hard to see that the new evidence at the Tanis site (e.g. Figure 7.1) is no silver bullet. If we look critically at the articles—both the online 'popular' article, and the scientific journal article it links to, DePalma et al. (2019)—they only provide strong evidence that many organisms perished when a large asteroid hit. The evidence found at the Tanis site in North Dakota does very little to support the central Alvarez claim, that an asteroid caused the K-Pg extinction, killing off all the world's non-avian dinosaurs (directly or indirectly). Almost 100 per cent of relevant scientists agree that an asteroid hit, and that it caused some chaos and killed lots of dinosaurs.[6] Whether it caused the End-Cretaceous mass extinction is another matter. It is obvious that the authors of the DePalma et al. (2019) paper are very much in favour of the Alvarez hypothesis, and indeed Walter Alvarez (of Alvarez et al. 1980) is one of the co-authors on the paper. But the paper's scientific content clearly has to do with the Tanis discovery, and an asteroid-impact explanation of the relevant empirical finds *at Tanis*. It has relatively little to do with explaining the mass extinction, although no doubt it adds incrementally to the evidence base.

Turning to the popular articles, these can be quickly dismissed as grounds for a justified belief. Certainly there has to be *quite* a strong feeling in the scientific community for so many 'popular' articles to state bluntly that an asteroid killed the dinosaurs: the writers of such articles usually know their science quite well, and it would damage their reputation if they were to write things that weren't true, or that weren't supported by many scientists. However, an 80 per cent consensus would be sufficient for these authors. After all, if there is an 80 per cent consensus regarding the Alvarez hypothesis, it will be very easy to find more than 20 'top experts' from around the world who strongly favour the Alvarez account. The way many of us view expertise, an agreement amongst 20 'top experts' would usually be more than enough for

[6] Note that the vast majority of sceptics vis-à-vis the Alvarez hypothesis think that the asteroid hit many years *before* (not after) the demise of the dinosaurs (e.g. Keller et al. 2004). Thus, just about all advocates *and sceptics* of the Alvarez hypothesis would agree that an asteroid hit the Earth and killed some dinosaurs.

us to strongly believe the proposed view. After all, we hardly consult 20 different doctors when we have an ailment: if we bother to consult three different doctors, and they all agree, and are all confident in their diagnosis, that would seem to be 'case-closed'. However, to base *scientific facts* on only 80 per cent consensus is a risky strategy, at least in the long term. The authors of those popular articles don't care too much about long-term risk: they just want to catch your attention with dramatic headlines. But in this book I really do care about long-term risk. Indeed, I want 'facts' to last forever.

Turning to the Wikipedia page, it is only as strong as its references. This brings us to Schulte et al. (2010). As already noted, in the abstract we find simply, '[T]he Chicxulub impact triggered the mass extinction.' More fully we find:

> The temporal match between the ejecta layer and the onset of the extinctions and the agreement of ecological patterns in the fossil record with modeled environmental perturbations (for example, darkness and cooling) lead us to conclude that the Chicxulub impact triggered the mass extinction.

In other words, it is being claimed that there is a 'match' between the timing of the Chicxulub impact and the extinction of the dinosaurs. It is also being claimed that models of what would have happened to the planet following the impact (ejection of material into the upper atmosphere blocking out sunlight, etc.) agree with patterns in the fossil record. Of course, dinosaurs on the other side of the world were not *directly* killed by the impact event; they were killed gradually by 'darkness and cooling'. Thus the authors use the word 'triggered' (see Section 5, below, for further discussion of the precise wording).

Now, 41 scientists is an extremely(!) tiny percentage of the relevant scientific community. Can we be sure that their opinion stands as a good proxy for 95 per cent of the community? A good way to check is to consider what might happen if in fact more than 5 per cent of the relevant scientific community wish to resist or reject the Alvarez hypothesis. We may suppose that, in that case, at least one *response article* to Schulte et al. (2010) would emerge in due course, probably quite quickly. After all, more than 5 per cent of relevant scientists is a very large number of scientists. That sub-community of scientists would be very keen to oppose Schulte et al. (2010) in print, presenting their reasons.

The existence of serious opposition to Schulte et al. (2010) is easy to establish. If one looks at the official article webpage for Schulte et al. (2010) one finds a section entitled 'Related Content', and there one can immediately see links to three letters: Archibald et al. (2010), Courtillot and Fluteau (2010), and Keller et al. (2010). The first of these explicitly objects to the suggestion of a consensus: 'The list of 41 authors, although suggesting a consensus, conspicuously lacked the names of researchers in the fields of terrestrial vertebrates, including dinosaurs, as well as freshwater vertebrates and invertebrates.' The second letter also objects: 'the Review does not give sufficient and accurate consideration to the volcanic hypothesis.' The third is even more explicit: 'To arrive at this conclusion, the authors used a selective review of data and interpretations by proponents of this viewpoint. They ignored the vast body of evidence inconsistent with their conclusion.' These three letters, in a leading journal, authored by more than 30 scientists from a wide range of countries, suggest that there definitely was not (in 2010) a solid consensus of the kind I am after. Instead, there seems to be scientific disagreement and debate. At this point, despite the very great difficulty of establishing (even roughly) the percentage of relevant scientists willing to call the asteroid story a 'fact', we might well assume that the percentage is something less than 95 per cent. We might reason that if the percentage really were above 95 per cent, it should be somewhat harder than this to find evidence of robust scientific debate, especially in such a leading journal.

But what about false balance? Is it possible that those three letters represent less than 5 per cent of the relevant scientific community, but the journal *Science* nevertheless wanted to represent both sides of the debate? The fact that there are *three* letters at first seems to put the lie to that idea—it gives the impression that *three different* groups of scientists independently wrote responses to Schulte et al. (2010). But are the three letters really independent? Is it possible that *one* group of scientists decided that the best strategy would be to separate their objections into three letters, as opposed to one? Consider the three letters:

1. 'Cretaceous Extinctions: Multiple Causes' (Archibald et al. 2010)
2. 'Cretaceous Extinctions: The Volcanic Hypothesis' (Courtillot and Fluteau 2010)
3. 'Cretaceous Extinctions: Evidence Overlooked' (Keller et al. 2010)

Vincent Courtillot is an author on letters 1 and 2, and Sunil Bajpai is an author on letters 1 and 3. Gerta Keller was only an author on one letter, letter 3, but she had co-authored with two of the authors on letter 1 just the year before, in 2009.[7] At the very least, it seems quite reasonable to suppose that the

[7] Keller, Sahni, and Bajpai (2009), in *Journal of Biosciences*; Keller et al. (2009), in *Earth and Planetary Science Letters*.

lead author of each of these letters knew of the other two letters. Perhaps they even shared drafts of the letters with each other, so as to make sure they weren't overlapping too much (since this would be a barrier to publication). A certain amount of coordination seems probable, meaning that the letters aren't as independent as they initially appear. If this did happen, does it matter very much? The worry is that one relatively small group of scientists might—with some ingenuity—be able to make it look like there is greater opposition to a scientific claim than there actually is. This thought might also be supported by the fact that the authors of these letters reference their own work quite a lot. For example, in the Courtillot letter, six out of 10 references are to Courtillot's own publications, and in the Keller letter, seven out of 11 references are to Keller's own publications.

The 'interested outsider', trying to navigate all this, may well feel unsure. On the one hand, there seems to be strong support for the Alvarez hypothesis. On the other hand, there is definitely some resistance to that hypothesis. And yet it is unclear how *much* resistance there is. We don't yet have a clear answer regarding the strength of the 'consensus', if that is even the right word. The existence of those three letters is *suggestive* of an opposition to the bald statement 'An asteroid killed the dinosaurs' that amounts to more than 5 per cent of the relevant scientific community. But since 5 per cent is a very large number of scientists (thousands), we are making quite a leap if we get to that conclusion from the existence of those three letters. We need to dig a bit deeper.

3. Assessing the Opposition—Second Pass

The editors of *Science* back in 2010 were placed in a difficult position. They actually received at least *seven* letters in opposition to the Schulte et al. (2010) article. In addition, the Keller et al. (2010) letter was originally significantly longer, including 27 references exemplifying the opposition literature (as opposed to the 11 references found in the published letter). Such information is usually lost to history, but thankfully in this case it was preserved in a May 2010 issue of *Geoscientist*.[8]

Clearly this increases the significance of the opposition, even if we search for ways to weaken it. For example, we might try to weaken it by identifying good

[8] Available at https://www.geolsoc.org.uk/Geoscientist/Archive/May-2010/ (last accessed February 2022).

reasons the editors of *Science* had for only publishing *three* of the seven submitted letters. For one thing, the three published letters already cover the three most obvious objections to Schulte et al. (2010): (i) there might be multiple causes for the K-Pg extinction, (ii) volcanism is too easily dismissed, and (iii) the evidence is selectively chosen. In addition, one of the letters—not chosen for publication—was written by Keller's former PhD student Sigal Abramovich, who had co-authored with Keller at least nine times between 2002 and 2009,[9] indicating a lack of independence between these authors. Another unpublished letter was authored by Ashok Sahni, but he is already a co-author on the published letter 1. Another possible way to attempt to weaken the opposition would be to point out that Keller et al.'s longer, original letter includes 27 references, but 15 of them are references to Keller's own publications, perhaps indicating a lack of diversity in the opposition literature.

But whatever way you look at it, the opposition is significant. And we can add to the strength of opposition back in 2010 a story concerning the strength of opposition *since* 2010. For example, we can find many articles published since 2010 in more-or-less-explicit opposition to an Alvarez consensus appearing in respected earth sciences journals, including:

(1) 'Cretaceous-Paleogene transition at the Paraíba Basin, Northeastern, Brazil: Carbon-isotope and mercury subsurface stratigraphies' (Nascimento-Silva et al. 2011, in *Journal of South American Earth Sciences*).

(2) 'Deccan volcanism linked to the Cretaceous-Tertiary boundary mass extinction: New evidence from ONGC wells in the Krishna-Godavari Basin' (Keller et al. 2011, in *Journal of the Geological Society of India*).

(3) 'Chicxulub impact spherules in the North Atlantic and Caribbean: age constraints and Cretaceous–Tertiary boundary hiatus' (Keller et al. 2013, in *Geological Magazine*).

(4) 'High-resolution Hg chemostratigraphy: A contribution to the distinction of chemical fingerprints of the Deccan volcanism and Cretaceous–Paleogene Boundary impact event' (Sial et al. 2014, in *Palaeogeography, Palaeoclimatology, Palaeoecology*).

[9] (1) Abramovich and Keller (2002), in *Palaeogeography, Palaeoclimatology, Palaeoecology*; (2) Abramovich et al. (2002), in *Marine Micropaleontology*; (3) Abramovich et al. (2003), in *Palaeogeography, Palaeoclimatology, Palaeoecology*; (4) Abramovich and Keller (2003), in *Marine Micropaleontology*; (5) Stuben et al. (2003), in *Palaeogeography, Palaeoclimatology, Palaeoecology*; (6) Abramovich et al. (2003), in *Marine Micropaleontology*; (7) Adatte et al. (2005), in *Bulletin de la Société Géologique de France*; (8) Keller and Abramovich (2009), in *Palaeogeography, Palaeoclimatology, Palaeoecology*; (9) Keller et al. (2009), in *Palaeogeography, Palaeoclimatology, Palaeoecology*.

(5) 'U-Pb geochronology of the Deccan Traps and relation to the end-Cretaceous mass extinction' (Schoene et al. 2015, in *Science*).
(6) 'End-Cretaceous extinction in Antarctica linked to both Deccan volcanism and meteorite impact via climate change' (Petersen et al. 2016, in *Nature Communications*).
(7) 'Mercury anomaly, Deccan volcanism, and the end-Cretaceous mass extinction' (Font et al. 2016, in *Geology*).
(8) 'End-Cretaceous akaganéite as a mineral marker of Deccan volcanism in the sedimentary record' (Font et al. 2017, in *Scientific Reports*).
(9) 'U-Pb constraints on pulsed eruption of the Deccan Traps across the end-Cretaceous mass extinction' (Schoene et al. 2019, in *Science*).
(10) 'Mercury linked to Deccan Traps volcanism, climate change and the end-Cretaceous mass extinction' (Keller et al. 2020, in *Global and Planetary Change*).

Some of these papers are more explicit on the K-Pg mass extinction than others. For example, in (3) we find, 'this [Chicxulub] impact predates the Cretaceous–Tertiary boundary by about 130–150 ka', and in (10) we find 'These results support Deccan volcanism as a primary driver of the end-Cretaceous mass extinction', whereas in (8) we find nothing more explicit than reference to 'evidence of volcanic halogen degassing and its potential role for the Cretaceous-Tertiary mass extinction'. In (6) we find simply: 'The cause of the end-Cretaceous (KPg) mass extinction is still debated.' Not all of these papers clearly or directly contradict the Alvarez hypothesis. But together they do suggest a debate that is very much 'open'.

Some of the same names keep cropping up in such articles. Gerta Keller and Thierry Adatte are major drivers of the resistance to the Alvarez hypothesis; they are each authors on six of the given ten articles (1)–(10). However, all in all, there have been a significant number of articles, with a significant number of authors, from a significant number of different countries, published in a significant number of different journals. To get published in all of these journals the authors would have had to get their articles past peer reviewers and editors who were *not* part of any in-group opposition to Schulte et al. (2010). This is especially true in the case of the publications in *Science* (Schoene et al. 2015; Schoene et al. 2019), a journal that Courtillot (1999, p. 138) claims was at one time biased towards the Alvarez hypothesis. It is also significant that there is minimal overlap between the authors of these ten articles (1)–(10) and the authors of those 2010 opposition letters in *Science*; in fact, the *only* overlapping names are Keller and Adatte. If we had found that all or most of the articles published since 2010 in opposition to Schulte et al. (2010) were authored by individuals named as authors on those three opposition letters, that might have been significant: it would indicate that a relatively small group of scientists was doing all the work to oppose the Alvarez claim. But that clearly isn't the case—there are seemingly plenty of scientists who were *not* involved in opposing the Schulte paper in 2010, and yet have *since* been involved in the serious-journal-article opposition to that paper. It's perhaps worth stressing that I'm not saying they oppose the Alvarez hypothesis; I'm saying they oppose the claim that the asteroid impact story is scientific fact. They might in fact think that the Alvarez hypothesis is the clear front-runner—that's very different from it reaching the status of 'fact'. It is also important to emphasise that the list of articles (1)–(10) is not the result of an exhaustive search of the literature. It is instead intended as an indication of what is out there, and to place the ball firmly in the court of anyone who would claim that 95 per cent of relevant scientists firmly believe the Alvarez hypothesis.

Apart from the published literature, there is another good way to judge the state of play in the community: conference activity. For example, we can look at the annual conference of the Geological Society of America (GSA) since 2001:

- At GSA2001 there was little obvious[10] discussion of the matter.
- At GSA2002 there was no significant discussion of the matter. In fact, the only reference to the K-Pg extinction in the conference presentation abstracts strongly favours the Alvarez hypothesis: 'A well known case in point is the Cretaceous-Tertiary boundary, where the discovery of an extraterrestrial signature, together with the presence of shocked minerals, led not only to the identification of an impact event as the cause of the end-Cretaceous mass extinction, but also to the discovery of a large buried impact structure about 200 km in diameter, the Chicxulub structure.'[11]
- At GSA2003 there was little obvious discussion of the matter.
- At GSA2004 there were three[12] sessions on the topic of 'Impact Geology', advocated by David King and Jared Morrow, with 30 presentations in all.

[10] Of course, there might have been substantial discussion informally—here I refer merely to the formal proceedings.

[11] See https://gsa.confex.com/gsa/2002AM/webprogram/Paper40726.html (last accessed February 2022).

[12] See https://gsa.confex.com/gsa/2004AM/webprogram/Session12885.html; https://gsa.confex.com/gsa/2004AM/webprogram/Session13328.html; and https://gsa.confex.com/gsa/2004AM/webprogram/Session13329.html (last accessed February 2022).

Whilst many of the presentations assumed or favoured the Alvarez hypothesis, some went against it, suggesting significant debate on the matter at this conference.

- At GSA2005 there was little obvious discussion of the matter.
- At GSA2006 there was a session on the topic of 'The Late Cretaceous-Early Tertiary Interval in the Atlantic Coastal Plain', advocated by William Gallagher and Kenneth Lacovara, with eight presentations.[13] There was another session on the topic of 'Extinction, Dwarfing and the Lilliput Effect', advocated by Richard Twitchett and Bridget Wade, with 13 presentations. Gerta Keller presented on the End-Cretaceous mass extinction in both of these sessions (opposing the Alvarez hypothesis). In yet a third session, on the topic of 'Terrestrial Impact Breccias', Thierry Adatte presented on the End-Cretaceous mass extinction, opposing the Alvarez hypothesis.
- At GSA2007 there was a session on 'Stratigraphy', advocated by Aditya Tyagi and Peter Holterhoff, with 15 presentations.[14] Adatte and Keller were once again involved. They were also involved—along with Silvia Gardin, Annachiara Bartolini, and Sunil Bajpai—in a poster session entitled 'Diversity, Evolution, and Turnover', with a paper entitled 'Age and Paleoenvironment of Deccan Volcanism and the K-T Mass Extinction'. Keller and Adatte were also involved in a session entitled 'Selectivity of Ancient and Modern Extinctions: Bridging the Gap Between Neontological Prediction and Paleontological Observation', with a paper entitled 'Main Deccan Volcanism Phase Ends at K-T Mass Extinction: Evidence from the Krishna-Godavari Basin, SE India'.
- At GSA2008 there was little obvious discussion of the topic.
- At GSA2009 there were three sessions on the topic of 'Volcanism, Impacts, Mass Extinctions, and Global Environmental Change', advocated by Keller, Adatte, and Wignall, with 39 presentations in all.[15] There must have been plenty of discussion of the Alvarez hypothesis; several scientists known to be opposed to Alvarez were either present, or were co-authors on the papers presented in the session. For example, in a paper involving Vincent Courtillot, Frédéric Fluteau, Anne-Lise Chenet,

[13] See https://gsa.confex.com/gsa/2006AM/webprogram/Session18052.html (last accessed February 2022).

[14] See https://gsa.confex.com/gsa/2007AM/webprogram/Session20127.html (last accessed February 2022).

[15] See https://gsa.confex.com/gsa/2009AM/webprogram/Session23746.html; https://gsa.confex.com/gsa/2009AM/webprogram/Session25172.html; and https://gsa.confex.com/gsa/2009AM/webprogram/Session25173.html (last accessed February 2022).

and Maud Moulin, the abstract states: 'The impact appears important but incremental, neither the sole nor main cause of the Cretaceous-Tertiary mass extinctions.'[16]

- At GSA2010 there was little obvious discussion of the matter.
- At GSA2011 there were two sessions devoted to 'Multidisciplinary Approaches to Studying the Causes and Consequences of Mass Extinction', advocated by Marc LaFlamme and Simon Darroch, with 27 presentations altogether.[17] Deccan Traps volcanism was debated in connection with the K-Pg mass extinction.
- At GSA2012 there was little obvious discussion of the matter.
- At GSA2013 there was little obvious discussion of the matter.
- At GSA2014 three sessions were devoted to 'Mass Extinctions', advocated by David Bond, Gerta Keller, and Thierry Adatte, with 35 presentations altogether.[18] Deccan Traps volcanism was debated in connection with the K-Pg mass extinction.
- At GSA2015 two sessions were devoted to the topic of 'Mass Extinction Causality', chaired by David Bond, Paul Wignall, and Mike Widdowson, with 39 presentations in all (including poster presentations).[19] The End Cretaceous mass extinction was debated, including the alternative explanation of Deccan Traps volcanism.
- At GSA2016 a session was devoted to the topic of 'Volcanism, Mass Extinctions, and Environmental Change', advocated by Thierry Adatte, Stephen Grasby, Gerta Keller, Blair Schoene, and David Bond, and involving 16 presentations.[20]
- At GSA 2017 two major sessions were devoted to the topic of 'Mass Extinctions', advocated by David Bond, Thierry Adatte, Gerta Keller, and Dougal Jerram, and involving 37 presentations in all (including posters).[21]
- At GSA2018 there was nothing obvious.

[16] See https://gsa.confex.com/gsa/2009AM/webprogram/Paper159950.html (last accessed February 2022).

[17] See https://gsa.confex.com/gsa/2011AM/webprogram/Session28725.html (last accessed February 2022).

[18] See https://gsa.confex.com/gsa/2014AM/webprogram/Session35318.html (last accessed February 2022).

[19] See https://gsa.confex.com/gsa/2015AM/webprogram/Session37770.html (last accessed February 2022).

[20] See https://gsa.confex.com/gsa/2016AM/webprogram/Session40122.html (last accessed February 2022).

[21] See https://gsa.confex.com/gsa/2017AM/meetingapp.cgi/Session/42879 (last accessed February 2022).

- At GSA2019 two major sessions were devoted to the topic of 'Volcanism, Impacts, and Phanerozoic Mass Extinctions', advocated by Paula Mateo, Gerta Keller, Stephen Grasby, Thierry Adatte, and David Bond, and involving 22 presentations in all.[22] Scientists known to be in opposition to Alvarez were heavily involved. Three other sessions at GSA2019 were entitled 'Unconventional Ideas and Outrageous Hypotheses',[23] but note that the volcanism explanation of the End-Cretaceous mass extinction would have enjoyed a serious hearing at the conference even without this 'Unconventional Ideas' session. Which is another way of saying that the volcanism theory is not considered an 'unconventional idea' by the community; it is *not* a 'fringe' theory.

So in 10 out of the last 19 GSA conferences the issue has been formally debated. None of this is to say that there aren't still a great many scientists willing to state unambiguously that an asteroid caused the mass extinction. For example, at GSA2018 one presentation includes in its abstract: 'This [Chicxulub] impact induced environmental perturbations on a global scale, resulting in a mass extinction event at the end of the Cretaceous period.'[24] And in general, GSA2018 seems to have been more in favour of the Alvarez hypothesis than other years (such as 2017 and 2019). Similarly when it comes to some of the other GSA conferences, such as GSA2002. In addition it is worth bearing in mind that not *all* papers presented at GSA conferences are ultimately considered 'serious': at GSA 2011 in Minneapolis, for example, a much-ridiculed paper was presented on a hypothesised giant prehistoric squid dubbed 'The Kraken'.[25]

But what stands out clearly from this brief sketch of 19 GSA conferences is how much the earth sciences community (in America, at least) actually *has* been willing to tolerate serious doubts about the Alvarez hypothesis. There is a stark distinction between the platform allowed by the community to doubt the Alvarez hypothesis, and the complete *lack* of a platform allowed by other scientific communities when it comes to any rare, lingering doubts about the

[22] For further details of the sessions and the presentations, see https://gsa.confex.com/gsa/2019AM/meetingapp.cgi/Session/47845 and https://gsa.confex.com/gsa/2019AM/meetingapp.cgi/Session/48658 (last accessed February 2022).

[23] See https://gsa.confex.com/gsa/2019AM/meetingapp.cgi/Session/47461 (last accessed February 2022).

[24] See https://gsa.confex.com/gsa/2018AM/meetingapp.cgi/Paper/323805 (last accessed February 2022).

[25] See e.g. 'The Giant, Prehistoric Squid That Ate Common Sense', available at https://www.nationalgeographic.com/science/phenomena/2011/10/10/the-giant-prehistoric-squid-that-ate-common-sense/ (last accessed February 2022).

30 examples listed back in Chapter 1. One interesting example concerns GSA2019, which included three sessions on the topic 'Unconventional Ideas and Outrageous Hypotheses', but which certainly blocked *some* hypotheses. In particular, the session received one submission from a climate change denier, but this was *not* allowed in the session.[26] There isn't a platform at a serious conference for a climate change denier, even in a session explicitly devoted to 'outrageous hypotheses'.

When it comes to the Alvarez hypothesis, the doubters might not have always had it easy, and Gerta Keller in particular has despaired of the community dynamics that have—sometimes—attempted to suppress (or at least inhibit) opposition to the Alvarez hypothesis. At GSA2019 Keller gave an invited talk entitled 'The Nasty 40-Year Mass Extinction Controversy', writing in the abstract:

> Scientists who questioned the [Alvarez] theory, revealed contrary evidence and proposed alternative causes were confronted with mass hostility, public ridicule, character assassination, prevention of publishing results and threats of destroying their careers. This tactic succeeded to silence almost all within just a few years and destroyed careers of the most prominent early opponents of the impact theory leaving just one and a small team to amass evidence and battle vicious attacks for the past three decades.[27]

Whilst there is no doubt plenty of truth in this—and it may remind us of the continental drift debate (recall Chapter 5)—formal academic debate on the topic has certainly been possible since 2004: as sketched above, the topic was formally debated at GSA conferences on several occasions both before and after the publication of Schulte et al. (2010). In addition to the GSA conferences it is also worth mentioning 'The Great GSL Chicxulub Debate' in 2004, hosted by The Geological Society of London.[28] This is yet another example of how the community has long been willing to tolerate debate concerning the cause of the K-Pg mass extinction. Keller may be referring to an earlier period, but if we ask after specific years for this period, when opposition to the Alvarez hypothesis was very significantly inhibited, it seems difficult to place. Alvarez et al. (1980) was influential, but hardly sufficient to bring about an 'instant

[26] Personal discussion with the organisers.

[27] For a full discussion, see 'The Nastiest Feud in Science', published in *The Atlantic* in September 2018, available at https://www.theatlantic.com/magazine/archive/2018/09/dinosaur-extinction-debate/565769/ (last accessed February 2022).

[28] See https://www.geolsoc.org.uk/chicxulub (last accessed February 2022).

consensus' in the first half of the 1980s; Courtillot (1999, p. 136) writes merely that, 'by the mid-1980s, the "impactists" had gained the upper hand.' The latter 1980s aren't a good candidate either; as Courtillot (1999, p. 56) notes, '[By 1986] A "volcanist" movement was already forming ranks against the "impactists".' In 1990 the Chicxulub crater was confirmed (see Courtillot 1999, p. 121ff.); was that sufficient to bring about a consensus? But no: in the years immediately following 1990 there was a process of dating the crater precisely (e.g. Swisher III et al. 1992), during which any announcement of certainty would be frowned upon as premature. Nor was there a solid consensus in 1997; for example, in Pope et al. (1997) we find, 'While the impact portion of the Alvarez hypothesis is now widely accepted, the second part, that of a causal link between the impact and mass extinction, remains controversial.' Such a statement would not make sense if it were really true that there remained only 'one and a small team' to fight against the 'truth' of the Alvarez hypothesis, as Keller claimed in her 2019 address to the GSA. Without dismissing her claims at all, we should be cautious about inferring that there was *such* a solid consensus on Alvarez that debate was almost shut down.

4. Should We Believe the Alvarez Hypothesis?

My focus here has been on the question of whether, judging by community opinion, we ought to believe the asteroid impact theory of the extinction of the dinosaurs. Significant community opposition suggests otherwise: there are indications that more than 5 per cent of the relevant community would resist the thought that the impact theory should now be referred to as a 'scientific fact'. The indications come from the publications in the literature, the activity at conferences, and the significant knee-jerk opposition to the Schulte et al. (2010) article in *Science*, which dared to hint at a possible consensus (although the actual word 'consensus' was introduced in opposition letter 1, not in the Schulte paper). Thus we reach the conclusion that it is premature to label the impact theory 'future-proof'.

It remains natural, however, to ask whether it is *reasonable to believe* in the Alvarez hypothesis. In many debates about scientific theories we are not really after future-proof scientific facts: we just want to know what it is reasonable to believe. Crucially, we are willing to accept that such beliefs are fallible. Suppose that we make a simple adjustment to my criteria for future-proof science, such that we only require a community consensus of 80 per cent. It seems reasonable to suppose that, if 95 per cent consensus is a sufficiently high bar for future-proof science, then 80 per cent is probably a sufficiently high bar for *justified belief*, where a certain (small) degree of fallibility is accepted. We might therefore ask: 'Is there (probably) an 80 per cent community consensus on the Alvarez hypothesis?'

Of course, reaching a number for the percentage that we are confident in requires a large-scale project. Ideally we would survey the opinions of several thousand relevant scientists (taking into account diversity requirements vis-à-vis background, gender, specialisation, etc.), and then we would extrapolate from there to reach an estimate. We would also have to fashion our survey question very carefully, such that it couldn't feasibly be misinterpreted (recall the discussion in Section 4 of Chapter 4). In practice, perhaps the best we could do is follow the example of previous such surveys of scientific communities, such as the surveys conducted regarding the theory of evolution, and anthropogenic climate change (e.g. by the Pew Research Center). Even more ideal would be a meta-analysis of several such surveys.

Absent such a large-scale study, what can we say? Certainly the majority scientific opinion still lies with the Alvarez hypothesis. Consider the following:

- Some of the relevant scientists are willing to state the Alvarez hypothesis as a fact in print (e.g. Chiarenza et al. 2020; Lyons et al. 2020). Within scientific communities this is very unusual, unless the evidence really is *very* strong. After all, 'scientists are a conservative and cautious bunch' (recall the discussion in Section 3 of Chapter 5). It is obviously very embarrassing to state that something is a fact, only to later be proven mistaken. Scientists' reputations really are at stake when they make confident pronouncements. And they don't *have* to make such pronouncements—nobody is pressurising them to do so, at least not in the Alvarez debate.
- The bulk of the literature lies with the Alvarez hypothesis. For example, there is a very substantial supporting supplement[29] to the Schulte et al. (2010) paper, containing 142 references from a very wide range of authors. The sheer *number* of references, as well as the *diversity* of authors, is something the opposition literature cannot compete with. This is also true of publications since 2010. I won't fully substantiate this claim here but, just to give a flavour, 10 very significant papers in

[29] See https://science.sciencemag.org/content/suppl/2010/03/02/327.5970.1214.DC1 (last accessed February 2022).

highly respected/top-tier journals explicitly supporting the impact hypothesis are:

- Renne et al. (2013), in *Science.*
  - Statement: 'Our results strengthen conclusions that the Chicxulub impact played an important role in the mass extinctions.'
- Vellekoop et al. (2014), in *PNAS.*
  - Statement: 'This impact winter [. . .] significantly reduced incoming solar radiation for decades. Therefore, this phase will have been a key contributory element in the extinctions of many biological clades, including the dinosaurs.'
- Brusatte et al. (2015), in *Biological Reviews.*
  - Statement: 'Given the weight of current evidence, we hold here that the bolide impact was probably the fundamental cause of the dinosaur extinction.'[30]
- Lowery et al. (2018), in *Nature.*
  - Statement: 'The Cretaceous/Palaeogene mass extinction eradicated 76 per cent of species on Earth. It was caused by the impact of an asteroid on the Yucatán carbonate platform in the southern Gulf of Mexico 66 million years ago, forming the Chicxulub impact crater.'
- Gulick et al. (2019), in *PNAS.*
  - Statement: 'These records of the K-Pg boundary offer critical insights into the environmental effects of the Chicxulub impact and connections to the global extinction event.'
- Henehan et al. (2019), in *PNAS.*
  - Statement: 'Our data suggest that impact, not volcanism, was key in driving end-Cretaceous mass extinction.'
- Chiarenza et al. (2020), in *PNAS.*
  - Statement: 'These results support the asteroid impact as the main driver of the non-avian dinosaur extinction.'
- Hull et al. (2020), in *Science.*
  - Statement: '[T]hese models support an impact-driven extinction.'
- Lyons et al. (2020), in *PNAS.*

[30] In Brusatte (2015)—a single-authored piece—we find an even clearer statement: 'To our surprise, our team of nearly a dozen dinosaur experts – often an argumentative bunch – came to a clear consensus: as popular wisdom has it, the extinction was abrupt, and an asteroid was primarily to blame' (p. 56). It is worth noting that this consensus of 'a dozen dinosaur experts' is totally *in*significant compared with the kind of 95% community consensus this book is concerned with.

        - Statement: 'An asteroid impact [...] led to the Cretaceous-Paleogene (K-Pg) mass extinction of 76 per cent species, including the nonavian dinosaurs.'
    - Goderis et al. (2021), in *Science Advances*.
        - Statement: 'The clear association of the Ir anomaly within the Chicxulub impact structure and the recorded biotic response confirms the direct relationship between the impact event and the K-Pg mass extinction.'
- In the opposition literature the same names crop up again and again, more so than in the pro-Alvarez literature. In the 10 publications just listed, favouring the Alvarez hypothesis, there are 143 distinct authors involved.[31] In the previous section I listed 10 publications (1)–(10) doubting—or at least expressing a certain degree of caution about—the Alvarez hypothesis, published between 2010 and 2020. But there are only 49 distinct authors involved, and Keller and Adatte are each involved in six of the 10 papers listed.
- Whilst the GSA conferences analysed in the previous section suggest an open debate, other conferences are far less balanced. For example, if we consider the Annual Meeting of the Palaeontological Association[32] over the same 19 years, 2001–19, we find considerable support for the impact story (sometimes simply stated, other times argued for), and no real opposition.[33]

We can't do much better than this, I believe, without undertaking a very substantial survey of community opinion. But we probably already have enough. Simply put, there is clearly very strong support for the impact hypothesis in the community, but the debate is still (narrowly) open. Whether it is *reasonable to believe* in the impact claim depends on how willing you are to embrace fallible beliefs. The more epistemically cautious person won't yet be comfortable with the impact claim—certainly not with the word 'fact'—but the moderate risk-taker will. Since there is a spectrum of attitudes

[31] This does *not* include all the drilling expedition scientists. For example, Goderis et al. (2021) includes within its author list 'IODP-ICDP Expedition 364 Scientists', but we needn't consider these extra names for present purposes.

[32] Comprehensive abstract booklets for all of these conferences can be accessed online at https://www.palass.org/meetings-events/annual-meeting (last accessed February 2022).

[33] A range of other conferences could also be considered, including the annual meetings of the European Geosciences Union/American Geophysical Union. I leave this as an exercise for the reader. It would be interesting, at least, if we could establish that the GSA is the *only* conference to tolerate serious doubts about the Alvarez hypothesis in the past 20 years. But, even if true, that wouldn't negate the main point: that the community *does* tolerate serious doubts about the Alvarez hypothesis.

to risk in any community (recall the discussion in Section 3 of Chapter 5), we shouldn't expect a yes/no answer to the question whether it is currently reasonable to believe the impact story. It's a personal decision.[34]

Since we aren't under pressure to actually make a decision, a good option is to sit tight and see what happens over the next 10 years. New and significant evidence has emerged in the 10 years spanning 2010–20, and we can certainly expect new and significant evidence to emerge during 2020–30. For example, there is probably *already* a 95 per cent community consensus that the Chicxulub impact did *not* happen as much as 300,000 years prior to the K-Pg mass extinction, as Keller claimed in a 2004 paper (Keller et al. 2004); Keller herself later adjusted the figure downwards, to '130–150 ka', in Keller et al. (2013). Renne et al. (2013) claim to show that the two events—asteroid impact and mass extinction—couldn't have happened more than 32,000 years apart. *If* the Chicxulub impact really did cause the K-Pg mass extinction, there is little doubt that this number will continue to drop over the next 10 years. We can watch and see if that happens.

Thus we might be optimistic that in a few decades—supposing the impact story is true—there *will* be 95 per cent community consensus, especially given the exponential growth of science[35] and the doubling rate of approximately 20 years (recall Section 3 of Chapter 2). However, a word of caution is in order. The Alvarez debate is a clear example of thoroughly interdisciplinary science. Consider the following passage from Courtillot's book *Evolutionary Catastrophes*:

> All branches of geology (including sedimentology, stratigraphy, paleontology, and mineralogy), of geochemistry and geochronology, of geophysics, but also of biochemistry, organic chemistry, nuclear physics, astrophysics, materials science, fluid and shock mechanics, many branches of applied mathematics and computer science, and oceanic and atmospheric sciences have made their contributions, and this list is certainly not complete! We can no longer remain unaware of the language of these disciplines, and the contributions they can make; yet at the same time it is impossible for any one person to master all of them. An uncomfortable position, this: being

[34] Or, better, it isn't a decision at all: it's a disposition one has no control over—nobody gets to choose how epistemically cautious they are.

[35] Cf. Brusatte et al. (2015, p. 629): 'Driving this dynamic period of research [the past 20 years] is an exponential increase in the rate of dinosaur discovery, with a new Mesozoic dinosaur species being named once every ~1.5 weeks at present.'

> condemned to reconcile extreme specialization (in one's own discipline) with a superficial but obligatory knowledge of most of the others.
>
> (Courtillot 1999, p. 155)

Uncomfortable indeed. The extreme specialist will be criticised for failing to see the wood for the trees. The jack-of-all-trades will be criticised for having such an embarrassingly superficial knowledge—by professional standards—of so many important scientific issues. The ideal situation, where an individual has a *very good* knowledge of *all* (or nearly all) of the issues, is a virtual impossibility. Perhaps occasionally a genius will emerge, who really does manage to have an excellent knowledge of several different disciplines bearing on the Alvarez debate (a true expert in biology, chemistry, physics, geology, and more!). But this would just be one individual, and we are interested in a community consensus involving hundreds of thousands of individuals. The opinion of the individual 'genius' carries very limited weight in the approach developed here.[36] And there is much to criticise when we consider an approach based on *identifying the geniuses*, seeing what they think, and inferring that that is probably correct. There are many obvious counterexamples to such an approach, some of which have already been discussed in previous chapters, including the classic quote from Lord Kelvin at the turn of the 20th century: 'There is nothing new to be discovered in physics now.'

Consider what this means for achieving 95 per cent community consensus of 'relevant scientists'. In a debate taking place within the confines of just *one* discipline, all (or nearly all) of the relevant scientists will have intimate knowledge of the data, the methods, the theory, and so on. But in a highly interdisciplinary debate many of the relevant scientists will have vast gaps in their knowledge. Consider the ultra-specialist in sedimentology: she will certainly have to be considered one of the relevant scientists in the Alvarez debate, because sedimentology bears heavily on that debate. But suppose she is a rank amateur in all of the other relevant fields. In that case she has one narrow perspective on an issue that demands multiple perspectives. We might say, 'That's fine, we are considering the *overall community view*, and this includes all different perspectives.' But the problem comes in making the transition from 'strong evidence' to 'scientific fact'. To reach 'scientific fact' I am requiring 95 per cent consensus, and yet a highly interdisciplinary context can be conducive to perpetual disagreement, *even when* the overall evidence is in fact incredibly strong. Consider: that ultra-specialist in

[36] Cf. Mercier and Sperber (2017), Chapter 18: 'Solitary Geniuses?'.

sedimentology might remain extremely cautious about using the word 'fact' simply because that word is *not* justified on the basis of her very narrow perspective. If there are enough ultra-specialists in that sort of position they will amount to more than 5 per cent of the relevant scientific community, and we will have to hold back on labelling the claim 'future-proof', according to my criteria. It is possible that in such circumstances we might *never* reach 95 per cent consensus, even if we would easily reach 95 per cent consensus if only, *per impossible*, all of the relevant scientists could see the 'full picture'.

Perhaps these could be grounds for lowering the 95 per cent bar to something more like 85 per cent *in highly interdisciplinary contexts*. One might claim that 85 per cent in a highly interdisciplinary context is just as impressive and hard-won as 95 per cent in another, simpler context. On the other hand, many of the epistemic concerns that have been raised throughout this book are just as serious, if not more so, in a highly interdisciplinary context. Consider the bandwagon effect, for example (recall Section 3 of Chapter 5): in an interdisciplinary context scientists in one discipline are probably *more* liable to 'follow the crowd' on certain matters in another discipline, given that they don't have the scientific expertise to assess things for themselves. At the very least, the bandwagon effect seems just as serious in an interdisciplinary context.

Thus given the choice between identifying future-proof science on an interdisciplinary matter by lowering the bar, and remaining sceptical (keeping the bar raised high, too high to reach), the latter may be preferable. Recall that I'm not here interested in cases where we absolutely *must* make a decision, one way or the other (e.g. are we going to ban CFCs, or not?). In those sorts of cases, a mere 51 per cent majority might sometimes be sufficient for action (depending on various factors). What I'm interested in here is identifying future-proof science, and the case of highly interdisciplinary science does *not* seem to me to be a case where we can confidently lower the bar. If the highly interdisciplinary nature of the Alvarez debate means that we will *never* reach 95 per cent agreement amongst relevant scientists, then so be it. However, I remain optimistic: important new empirical data, with the capacity to influence, is being published all the time. The new 'Tanis' site (DePalma et al. 2019) shows the kind of amazing empirical details it is possible to discover. There are surely many more such discoveries out there waiting to be made, of great significance for the debate. In addition, there are other cases in science where a 95 per cent consensus *has* been reached in a highly interdisciplinary context (e.g. anthropogenic climate change). Thus in my personal view, when it comes to the Alvarez proposal, the truth will out.

5. Coda on Approximate Truth

Back in Chapter 1, I stated: 'I don't deny that there will be *adjustments* to scientific ideas in the future. Just about any scientific idea one can imagine will be subject to some kind of refinement over the next tens/hundreds of years', and 'If we turn back to the concept of "future-proof science", then, I do want this to be compatible with adjustments.' So instead of asking whether the bold Alvarez statement is absolutely set in stone, we should be asking whether it is future-proof *modulo minor adjustments.*

What would a 'minor adjustment' of the Alvarez hypothesis look like? Perhaps that it only *partially* caused the K-Pg extinction? Something that is widely rejected in the community is the idea that the K-Pg extinction was *causally overdetermined*, such that it would have happened anyway, even without the asteroid. After all, it is unthinkable that there were *two* events, happening at the time same, each *individually* sufficient to cause the very same mass extinction. An interesting intermediate option is that a *very slow* mass extinction was already happening 66mya, due to Deccan Traps volcanism, and then an asteroid came along and turned it into a much quicker mass extinction. However, in this scenario it would still be the case that the asteroid caused *the* K-Pg mass extinction; without the asteroid it would have been a very different extinction event.

A closely related option is that the asteroid caused a dramatic extinction event only because it was helped along by other causal factors. Suppose, for example, that massive volcanism had partially poisoned the atmosphere for a great many years, such that organisms all around the world were struggling to survive year on year. In this scenario, one might say that the asteroid and the volcanism were 'jointly responsible' (something like 50:50): without the asteroid there wouldn't have been anything like the K-Pg mass extinction, but in addition *without the volcanism* there wouldn't have been anything like the K-Pg mass extinction. Imagine for a moment the very same asteroid hitting a planet where atmospheric conditions were *ideal* for life. In that case, perhaps many of the dinosaurs would have made it through the impact winter engendered by the impact. Courtillot (1999, p. 142) writes, 'Must we allow that on a single occasion, while the Deccan Traps were already erupting and the biosphere was sorely put to the test, an impact occurred at the same time, dealing a further blow to species so severely tried already? This is what seems most likely today.'

Consider now the *extent to which* the biosphere was 'sorely put to the test' by volcanism. This is clearly a matter of degree. If it was only 'put to the test' a very little bit, then the asteroid would be by far the most significant causal factor. If it was put to the test *a lot*, then it is possible that the asteroid was *a* causal factor, but only something like 50 per cent (as opposed to nearly 100%). Now we must ask, under which of the variety of possible scenarios would it be reasonable to say that the Alvarez hypothesis is *approximately true*? Clearly if the asteroid is a hugely significant causal factor, then the hypothesis is approximately true, if not plain true. But if it is more like 50 per cent, what should we say?

Personally, I don't think there is any deep problem here. I don't think we need a 'theory of approximate truth' to navigate our way through this. In practice, people know the difference between clear cases of 'minor adjustment', and clear cases of 'major adjustment'. Claims shown to require a minor adjustment at a later time were approximately true, and claims shown to require a major adjustment were not. Of course, there will be plenty of intermediate cases, but there are vague cases in any distinction (without destroying the distinction). In the case of the Alvarez hypothesis, if it turns out that the asteroid impact was only *partly* responsible for the K-Pg extinction, then the claim was approximately true in a case where it was very significantly responsible. We don't need to quantify 'very significantly responsible', since we are dealing with vagueness, and any attempt at quantifying matters would be inappropriate. If it turns out that the species of the world were already on their knees 66mya due to volcanism, and the asteroid played the rather minor role of pushing them over the edge (the straw that broke the camel's back), then it would *not* be reasonable to say that the Alvarez hypothesis was approximately true.

An important consideration in all this is the widely held assumption that any 'impact winter' caused by the Chicxulub asteroid would not have been survivable for non-avian dinosaurs. Crucially, the assumption is that it wouldn't have been survivable *even without any volcanism*. Models of the Chicxulub impact winter support this thought. For example, the models discussed and evidenced in Vellekoop et al. (2014), Chiarenza et al. (2020), and Lyons et al. (2020) don't leave much room for a mere 50 per cent causal role for the asteroid. Thus it seems very unlikely that issues of approximate truth, such as those sketched above, will ultimately play a major role in this debate. Much more likely, either the asteroid was a highly significant causal factor, or it wasn't involved at all. Researchers favouring the Alvarez hypothesis usually make room for the possibility that the asteroid was not 100 per cent responsible for the K-Pg extinction. For example, in Brusatte et al. (2015) we find reference to a 'key role' for the asteroid, as the 'fundamental cause of the extinction' (p. 639). The authors are certainly *not* saying that volcanism contributed nothing at all. Brusatte (2015) states explicitly: '[T]he asteroid was not the whole story. The big plant-eating dinosaurs did undergo a bit of a decline right at the end of the Cretaceous' (p. 58). But it was only *some* dinosaurs, and it was only a *bit* of a decline. Brusatte et al. (2015) dismiss the idea that volcanism was a *major* causal factor, writing, 'There is no evidence for a global, long-term decline in the diversity of non-avian dinosaurs prior to their extinction' (p. 639).[37]

6. Concluding Thoughts

In conclusion, the asteroid impact account of the demise of the dinosaurs shouldn't yet be claimed to be future-proof. Rather significant debate continues within the earth sciences community, both in print and also at major conferences such as the GSA. We also need to bear in mind a large subsection of the earth sciences community that doesn't work on the K-Pg extinction, and would never publish a paper contributing to the debate either way (they are still 'relevant scientists', though, because they work in the right field, publish in the right journals, attend the right conferences, and so on). Many such individuals aren't providing any resistance to the Alvarez hypothesis, but they also cannot be considered to contribute to a 95 per cent consensus. Perhaps if directly asked about it, they would say 'I'm not sure, I don't work on that issue.' They might also say 'My hunch is that the impact story is right, but I wouldn't be comfortable calling it a "scientific fact".' Either of these responses would mean that they would *not* contribute towards a solid consensus of the kind this book is concerned with. The existence of such on-the-fence or 'lukewarm' individuals makes attaining a 95 per cent consensus particularly hard, meaning that such a consensus, when it exists, is especially hard won, and therefore especially significant.

My impression is that we are perhaps moving slowly towards 95 per cent consensus. The epistemic situation the human race finds itself in right now vis-à-vis the Alvarez hypothesis—strong community support but without a solid consensus, and with *some* continuing debate at relevant conferences and in respected journals—is a stage we passed through on the way to *all* of the 30 examples of future-proof science listed in Chapter 1. In addition, there are very

[37] A final thought is that the asteroid impact induced 'accelerated volcanism' (Renne et al. 2015). But in this case it is clear enough that the asteroid was very significantly responsible (the 'trigger'), even if volcanism was also a significant factor.

few examples of us reaching this epistemic stage, without then progressing to 95 per cent consensus. However, we must keep reminding ourselves that we are not there yet. Brusatte et al. (2015, p. 629) use the interesting term 'emerging consensus', referring to the status of the Alvarez hypothesis. This is a curious term, since it allows that we haven't yet reached a (solid) consensus, and yet it seems to predict that one is on the way. Whilst I am sympathetic to this, I think we would do better to avoid to word 'consensus' entirely until we have really reached something like 95 per cent consensus (with all the caveats detailed back in Chapter 5). In the meantime we can say something regarding the strength of opinion *right now*, as opposed to trying to predict what is 'emerging'.

If we *are* heading towards 95 per cent consensus, how soon we might get there depends on the research that takes place in the coming years, including the discoveries that are made and the techniques that are developed. We have to accept that it is a slow process: the 40 years since the Alvarez paper of 1980 have been intense, and much has been achieved, but we are not there yet. It is worth considering the importance of Gerta Keller's project over the past few decades. Whether or not we reach 95 per cent consensus on the impact story in the next few decades, Keller's project has been incredibly important for the debate. In the face of her criticisms and objections, the impactists have been forced to work *very* hard to close every possible loophole in their arguments, gather the right kinds of data, improve their techniques, and so on. We can all think critically, and the impactists can certainly think critically about their own assumptions and inferences. But the ultimate critical thinker is the person who *really thinks the ideas are wrong*. If scientists want to be pushed as hard as they can be pushed to refine their arguments and really make their case, then they *want* a serious opposition community. There is no doubt some truth in Keller's claim that this has been a 'nasty 40-year controversy', and I would hardly hold up the Alvarez debate as an exemplar of science working at its best. But at the same time, rather heated debate can sometimes be a good way to ensure that both sides dig deep and find the best possible arguments, both *for* their own view, and *against* the opposing view(s).

One might suppose that there will *always* have to be many years—even decades—of intense debate, whenever a new scientific claim is put forward. After all, there has to be international scrutiny—every step of the argument needs checking—and we may need many iterations of research before the community starts to feel confident, and a consensus forms. But this isn't quite right. Sometimes a scientific consensus can form in a matter of a few *weeks*! Just such a case is investigated in the next chapter.

8

Scientific Knowledge in a Pandemic

1. Misuse and Abuse of 'Scientific Consensus'

Showing that we ought *not* to believe that some scientific claim is future-proof is usually easy. All we need do is bring to bear the criteria for future-proof science, and find a criterion that isn't met. And, since the criteria are rather stringent, it is quite easy for a contemporary scientific claim to fail to meet all of the criteria. As discussed in Chapter 7, even a claim as strongly favoured as the Alvarez hypothesis probably falls short of meeting the criteria. This needn't be a problematic result, of course: there *does* remain (some) genuine scientific debate concerning the Alvarez hypothesis and so, despite the strong evidence in its favour, it is a bit too soon to declare it a future-proof scientific fact.

Notice though how easily one could be led to believe that there *is* a solid scientific consensus regarding the Alvarez hypothesis. It can take quite a bit of work (as seen in Chapter 7) to work out whether there is a *mere majority of opinion* regarding the truth of a claim (e.g. 75%) or whether there is a solid consensus (95%) of the kind that, I argue, is sufficient for future-proof science. After all, in both cases it is relatively easy to find many scientists (thousands!) who favour the proposition in question.

The difficulty one can have deciding between these possibilities—mere majority, or solid consensus—is often at play in contemporary debates, and can be very confusing for the public. The Covid-19 coronavirus pandemic is an important case in point, since some claims about the illness and its cause enjoyed *mere majority* support from relevant scientists, whilst others enjoyed a solid consensus. For example, in the UK, one thing folk desperately wanted to know was whether to prefer a full country-wide lockdown, or else a more targeted approach focusing on certain regions and especially vulnerable sectors of society. At the end of September 2020 it seemed to many that there was a solid scientific consensus regarding this matter. For example, in a *British Medical Journal* 'opinion' letter,[1] 40 relevant experts (scientists and

[1] See the letter at https://blogs.bmj.com/bmj/2020/09/21/covid-19-an-open-letter-to-the-uks-chief-medical-officers/#comment-5,079,217,714 (last accessed February 2022).

Identifying Future-Proof Science. Peter Vickers, Oxford University Press. © Peter Vickers 2023.
DOI: 10.1093/oso/9780192862730.003.0008

practitioners) urged the government to 'suppress the virus across the entire population'.[2] And soon afterwards a *Guardian* article[3] (dated 29 September 2020) urged that 'The overwhelming scientific consensus still lies with a general lockdown.' One may now make a simple inference: if there is an overwhelming scientific consensus that we need a general lockdown, then we need a general lockdown! However, a statement that there is an overwhelming scientific consensus falls very short of establishing that there *is* an overwhelming scientific consensus. Little doubt, the author of that *Guardian* article—the respected science writer and former researcher in immunology Stephen Buranyi—actually thought that there *was* an overwhelming scientific consensus in favour of a general lockdown. But in hindsight it seems clear that that wasn't the case, at least not at the time. On 3 October 2020 a letter[4] signed by 66 GPs was sent to the government, stating, 'We are concerned due to mounting data and real world experience, that the one-track response threatens more lives and livelihoods than Covid-lives saved.' In other words, they were against a general lockdown.

Folk at the time were, naturally, desperate for a clear way to proceed. Thus—in a time of great uncertainty—many were looking for something solid to hold onto and use to make sensible decisions. A solid scientific consensus in favour of one way forward would be just such a solid platform, whereas an admission that the scientific community was divided on the issue would be simply unhelpful in the face of the urgent need for action. There can be a tendency, then, to find scientific consensus when really it isn't there to be had, and this is surely what Stephen Buranyi did when he penned that *Guardian* article.

But, in addition, it is crucial to note that whether it is better to go into a general lockdown, or follow a targeted approach, *is not a scientific question*. It depends on values, and thus takes us into the realm of philosophy. For example, *science* cannot tell us whether it is worse for one (additional) person to die of Covid, or for 10,000 (additional) people to lose their jobs. As I have

[2] Here I follow the discussion in 'Coronavirus: Is the cure worse than the disease? The most divisive question of 2020', *The Conversation*, 6 October 2020, available at https://theconversation.com/coronavirus-is-the-cure-worse-than-the-disease-the-most-divisive-question-of-2020-147343 (last accessed February 2022).

[3] 'Talk of a scientific rift is a dangerous distraction in the fight against Covid-19', *The Guardian*, 29 September 2020, available at https://www.theguardian.com/commentisfree/2020/sep/29/rival-scientists-lockdowns-scientific-covid-19?ref=hvper.com&utm_source=hvper.com&utm_medium=website (last accessed February 2022).

[4] To see the letter, follow the link in the *The Conversation* article 'Coronavirus: Is the cure worse than the disease?'

discussed elsewhere,[5] we must observe a crucial distinction between the concept of a scientific consensus *where you merely mean that the vast majority of scientists agree*, and the concept of a scientific consensus *where properly scientific activity has led to community agreement*. In this book, when I talk about the criteria for future-proof science, it is always the latter I have in mind.

2. When Was the Cause of Covid-19 Known with Certainty?

The issue of how easy/hard it is for a scientific claim to meet the criteria for future-proof science is a matter of crucial importance. Setting the bar so high makes it very hard for a claim to come out as a definite future-proof scientific fact. This is fine for the Alvarez hypothesis, but for other scientific claims there are very serious consequences if we are only 'quite sure'. Suppose we are only 'quite sure' that the leading vaccine being put forward for Covid-19 is safe. Or suppose we are only *almost* certain that Covid is caused by a novel coronavirus.

In fact, when we think about the coronavirus pandemic we have a potentially serious threat to the position put forward in this book. My criteria might suggest that for something to be confidently labelled a future-proof scientific fact requires years of international scrutiny. But in 2020, during the coronavirus pandemic, any scientific claims regarding the virus, or the illness, couldn't possibly have been thoroughly scrutinised by the community over a long period of time. In which case we are dangerously close to saying that scientists in 2020 didn't know *anything*, for sure, regarding the illness and its cause. But that seems clearly wrong. It is compatible with the claim that they didn't know *for sure* that a virus was responsible, and that they couldn't *definitely* rule out 5G (the fifth generation technology standard for cellular networks) as the cause.

But recall what was said in Chapter 5, regarding how some theoretical claims have now been 'proven by observation'. In the case of continental drift we have gone on a 100-year journey from (i) a tentative hypothesis based on scant evidence, to (ii) a definite theory based on good evidence, to (iii) direct (or, more accurately, sufficiently-close-to-direct) observation. When it comes to using *observation* as grounds for believing a scientific

[5] I originally discussed this distinction in an article for the Institute of Arts and Ideas (IAI), available at https://iai.tv/articles/idealism-panpsychism-and-science-auid-1700 (last accessed February 2022). See Miller (2013) for important and relevant discussion concerning consensuses that have come about in different ways.

claim, there is no need to wait for many years of international scrutiny. A scientific claim can be regarded as future-proof whenever the thing claimed has been sufficiently-directly observed. In practice, any claim to 'direct observation' will of course need to be scrutinised by the community, but that process needn't take *years*. In some cases, it might take only a few months. Perhaps even a few *weeks*. The case of SARS-CoV-2 and Covid-19 is a case in point.

2.1 Kinds and Outliers

When it comes to the Covid-19 coronavirus pandemic, which scientific claims could be verified by observation? Could scientists actually *observe* the causal relationship between a novel coronavirus and the Covid-19 illness? Stretching the common-sense notion of 'observation' just a little bit, I think it's reasonable to say that they could, and even quite early on. For example, patients suffering from the new illness could have various cells analysed, and it could be established that those cells had been taken over by a novel virus. The presence of sufficient virions accumulating in and around the cells of an infected person is a reasonably good sign that the virus is the cause. As already noted several times in this book, however, 'scientists are a conservative and cautious bunch' (Boudry 2020). In the early stages of analysing a new illness, scientists are cautious about asserting a causal link between a novel virus and the illness, *even when* they have solid evidence of a novel virus in/around the cells of the patients. Consider, for example, Drosten et al. (2003), a paper published just a few months after the first cases of SARS were identified in 2003. The study in question clearly established the existence of a novel coronavirus in SARS patients. The authors write:

> The high rate of positivity among patients with probable cases during an outbreak of SARS in Hanoi, in conjunction with the complete negativity among all healthy contacts of patients affected by the same outbreak, provides evidence of an association between the disease and the presence of this novel virus.

And yet, they are fully aware that sometimes evidence can be misleading. They continue:

> One should bear in mind, however, that in the past, viruses have been initially isolated from patients with a specific disease but subsequent investigations revealed no actual association at all.

They then cite a paper by Kew and Kassianides (1996), which theorises that a new 'Hepatitis G virus' in a certain set of patients was actually an 'innocent passenger'. Thus mere correlation—even a strong correlation—is not enough, and early on in the SARS case scientists were only willing to conclude, 'The novel coronavirus might have a role in causing SARS' (Drosten et al. 2003, p. 1967).

One must also bear in mind that SARS was the first serious case of a coronavirus infecting the human population. Whilst scientists had been working on coronaviruses for decades, and two human coronaviruses had already been discovered in the 1960s—HCoV-229E and HCoV-OC43—SARS was significantly different. By contrast, when MERS came along in 2012, scientists could draw on all of their experience with SARS. Consider, for example, one of the early MERS papers:

> The clinical picture was remarkably similar to that of the severe acute respiratory syndrome (SARS) outbreak in 2003 and reminds us that animal coronaviruses can cause severe disease in humans.
>
> (Zaki et al. 2012, p. 1814)

In the MERS case, very high confidence that the novel coronavirus was causally responsible for the new respiratory illness came sooner than with SARS. Zaki et al. (2012) were able to make use of a method for detection of 'all known and possibly unknown coronaviruses' developed in the wake of the SARS epidemic (Vijgen et al. 2008). Indeed, the method *was* able to detect the new MERS coronavirus: MERS-CoV was sufficiently similar to previously identified coronaviruses that Vijgen et al.'s 'pancoronavirus' method caught it.

There is an important point to make here about kinds, especially biological kinds. Biologists are used to outliers and exceptions. For example, just because all ant species identified to date live in colonies, and perceive smells with their antennae, doesn't make it a certainty that a new species of ant will do the same. Take the case of the discovery of the first vegetarian spider (Meehan and Curry 2008). At the time of the discovery in 2008, more than 40,000 species of spider were documented, and all of them were carnivorous. And yet, upon discovering a new species of spider, one couldn't say *for sure* that it would be carnivorous, even with such a solid base of knowledge to draw from. And the same goes for any other characteristic one might consider. Indeed, sometimes it is an outlier species that one expects to discover next—all of the 'easy' examples have already been found. New species of bat are sometimes discovered in extremely remote caves, for example.

Sometimes the scientific community is caught out: the scientists develop very strong expectations regarding a kind, based on a wealth of experience, and these expectations cause trouble when they are faced with a radical outlier. This is what happened in the case of giant viruses, first observed in the early 1990s, but only identified *as* viruses 10 years later. In the early 1990s virologists had a huge wealth of experience with viruses, and all of the known viruses at that point were very small compared with bacteria. In addition to this empirical knowledge, it was theorised that being so small was hardly accidental—viruses supposedly *had* to be so small. As Cleland (2019, p. 182) relates:

> Their small size was explained theoretically in terms of their greater functional simplicity. Viruses reproduce by commandeering the genetic and metabolic machinery of a host cell, and hence do not need the complex molecular machinery (ribosomes, cell membranes, cell walls, etc.) of bacteria, which are self-reproducing. They consist of little more than protein coated DNA or RNA molecules. The "particles" discovered inside amoebae were huge compared to known viruses, with a total diameter of approximately 750 nm (including fibers extending from their capsids), larger than some parasitic bacteria. In addition, they stained gram-positive, reinforcing the notion that they were bacteria.

Thus even with a huge number of prior examples of the kind, and theoretical reasons to suppose that *all* members of the kind (in this case 'virus') ought to share a certain property, a certain degree of caution is to be recommended. Unknown instances of the kind might yet differ in surprising ways.

Apply this now to species of coronavirus. Undoubtedly there are many species of coronavirus not yet discovered. Indeed, we are sure that there are many species of bat not yet discovered, and new coronavirus strains are commonly found in bat populations. Extreme outlier bat species are likely to exist in extremely remote caves, and it wouldn't be surprising if highly unusual coronavirus species were to be found in such outlier bat species. Thus the method of Vijgen et al. (2008), based as it is on currently known coronaviruses, probably does *not* work as a method for identifying literally *every* species of coronavirus that does exist, or could ever exist.

Thus whilst scientists learned much of value from the SARS and MERS epidemics, they couldn't immediately say *too* much with absolute confidence about the new coronavirus SARS-CoV-2. Even when they knew it was a novel coronavirus, they needed to be cautious regarding both (i) whether it was causing the new illness, and (ii) how exactly it behaved inside the body. Even something common to all other known coronaviruses—protruding virion spikes are used as keys to gain access to host cells—could not be asserted as a fact; SARS-CoV-2 could yet be a highly unusual species, unlikely as this was.

With such caution assumed, we should ask again: How soon did scientists know for sure that SARS-CoV-2 was the cause of Covid-19? Turning to the literature, they were sure within a matter of a few *weeks* (e.g. Tan et al. 2020; Huang, C et al. 2020). Squaring this with the message at the heart of this book is not easy. This book urges caution unless there has been extensive peer scrutiny, and this most obviously takes place over a period of several *years*. We must ask: How could scientists be *so* sure, so soon, regarding the connection between SARS-CoV-2 and Covid-19? For example, Tan et al. (2020) and Huang, C et al. (2020) were both originally published in January 2020, a matter of weeks after the first cases of the illness. Huang, C et al. (2020) states explicitly: 'A recent cluster of pneumonia cases in Wuhan, China, was caused by a novel betacoronavirus, the 2019 novel coronavirus (2019-nCoV).' There is no doubt about it, and the only thing we would now change about this statement, looking back, is the name of the virus.

2.2 The Empirical Route to Future-Proof Science

Suppose—as seems right—that scientists *were* certain, within a mere matter of a few weeks, that the novel coronavirus was the cause of the new respiratory illness. Is it reasonable to say that they established this fact through direct observation? This might initially be resisted. As already noted, the mere *presence* of a novel coronavirus in/around the cells of Covid patients could be quickly established, but couldn't they yet be 'innocent passengers' (to use the term of Kew and Kassianides 1996)? Would it be outrageous to suggest that an unknown causal factor both causes Covid and also somehow leads to the emergence of SARS-CoV-2 from some kind of dormancy (a common-cause explanation of the correlation between SARS-CoV-2 and Covid)? A sceptic might agree that the standard scientific account is probably the most likely explanation (SARS-CoV-2 causes Covid), whilst insisting that we are far short of labelling that story 'future-proof'. Several years of scrutiny are required for that degree of confidence, suggests this cautious protagonist.

The cautious protagonist will claim the epistemic high ground here. Those scientists who stated within a few weeks that they *knew* that SARS-CoV-2 is responsible for Covid were sticking their epistemic necks out, and they *might* have been wrong. Whereas, it is impossible for the sceptic to be wrong—after all, the sceptic is merely telling us to *take our time*, to *wait* for more and better evidence before we state that it is a fact.

The case of SARS-CoV-2 and Covid-19 is a good example of how the cautious sceptic *can* 'get it wrong', in a certain sense. Relevant here is Hendry's (2020) point that '[e]pistemic rationality requires us to balance the avoidance of error with the possibility of missing out on truth' (p. 482). As already discussed back in Chapter 1, extreme caution regarding the link between CFCs and ozone depletion would have led to a delay in acting, and that delay would have been a death sentence—at the hand of skin cancer—for many individuals who are alive today. How we handle doubt and knowledge in the coronavirus pandemic has similar high stakes: too much caution is dangerous.

How did the scientists know so soon, then? First of all it must be acknowledged that they were drawing on a huge wealth of both theoretical and empirical knowledge. The empirical knowledge in particular is striking. Reliable observations of coronaviruses were already being achieved in the 1940s (Figure 8.1a), 1950s (Figure 8.1b), and 1960s (Figure 8.1c). By the mid-1960s coronavirus virions could be observed multiplying within host cells. A virologist with years of experience studying cells can see, directly, that the virus is multiplying inside the cell (Figures 8.2a, 8.2b). In the decades that followed, the scientific community built up an extraordinarily rich picture of how viruses generally, and coronaviruses in particular, cause illness.

Fast-forwarding to the SARS and MERS epidemics, electron-microscope techniques led to very detailed images of the behaviour of the virus. And when SARS-CoV-2 came along in 2019, scientists were able to find and image the virus even as early as December 2019. Given all their knowledge of SARS and MERS, there couldn't be any reasonable doubt that 'SARS-CoV-2 + Covid-19' was a very similar case to what had come before. In other words, the new case was no 'outlier' (Figures 8.3, 8.4).

Special moments in the behaviour of viruses can also be captured with modern microscopy techniques. For example, one can see the precise moment that an individual virion ejects its genetic material (Figure 8.5). In another case, the precise moment a virion infects a host cell can be captured by observational techniques (Figure 8.6).[6]

[6] Figure 8.6 shows the precise moment when a T7 virion infects an *E. coli* cell. The authors provide many images, and even videos, showing how these particular virions (from the family *podoviridae*) find a suitable access point on the surface of the host cell, latch on, and infect cells using a 'tail' (Figure 8.6E) which reaches down into the cell's cytoplasm.

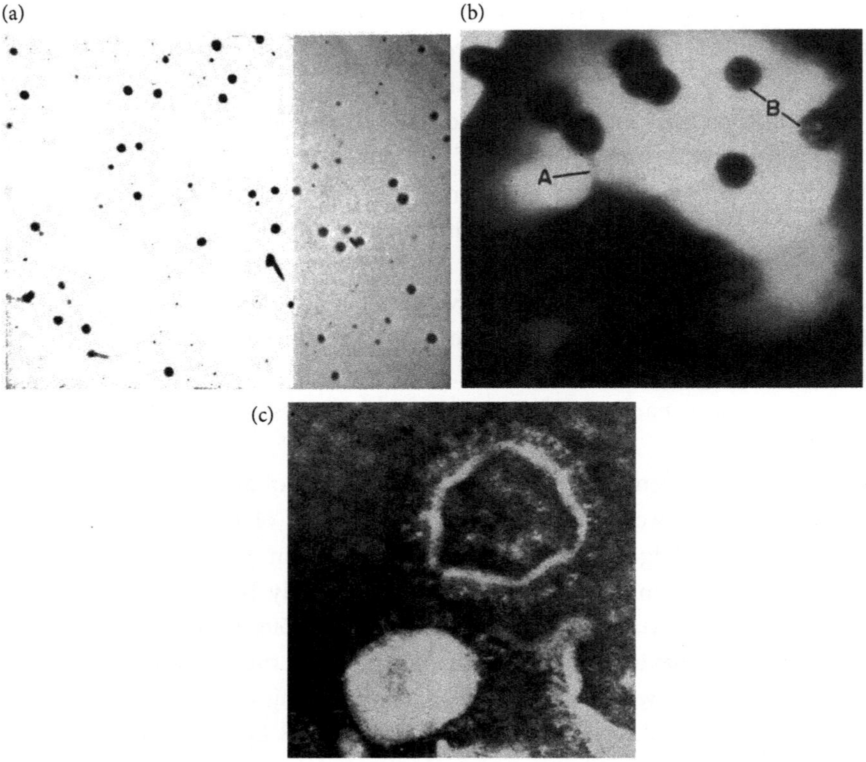

Figure 8.1 Direct images of coronavirus taken from: (a) Reagan *et al.* (1948); (b) Domermuth and Edwards (1957); (c) Berry *et al.* (1964).

Of course, scientists do not merely access virions and virus behaviour by observational techniques. They employ a wide variety of different techniques, many of which are far more *in*direct and based heavily on theory, including reverse-transcription polymerase chain reaction, or RT-PCR (e.g. Wang et al. 2020). Observation plays a crucial role, however, in testing the reliability of the indirect approaches. Various indirect approaches test for coronaviruses, but actually seeing one with an electron microscope is an important moment, epistemically speaking.

Hacking (1983, Chapter 11) asks 'Do we see with a microscope?', and there has been plenty of literature on this topic. Admittedly we don't *directly* see coronavirus virions with an electron microscope. But any scientific debate concerning whether we are really observing coronavirus virions when we see images such as those in Figure 8.4 was closed down many years ago; there hasn't been any scientific doubt on this issue for a long time. It would be outrageous for the sceptic to look at Figure 8.4 and say, 'One day we'll

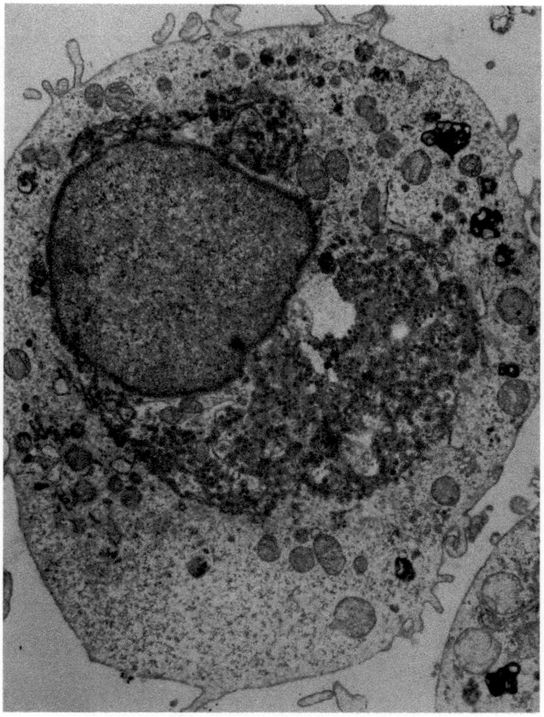

Figure 8.2a Direct image of a cell from a mouse liver, infected with coronavirus, 20 hours after inoculation

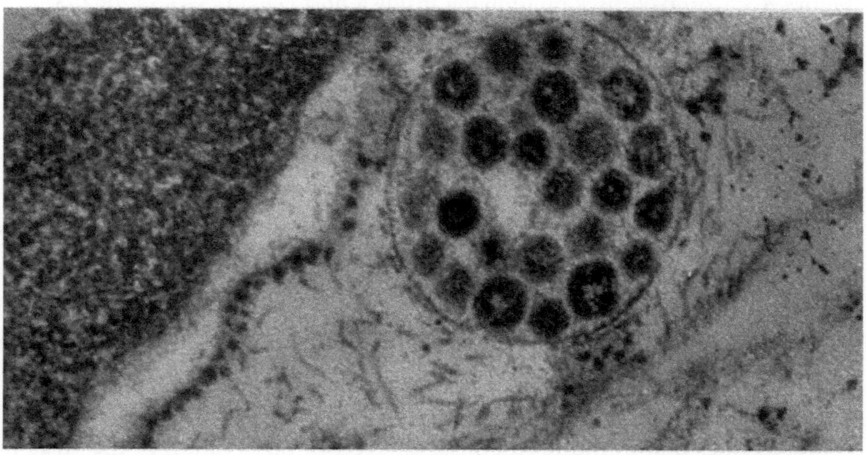

Figure 8.2b Direct image of coronavirus virions accumulating within a human cell, 12 hours post-infection

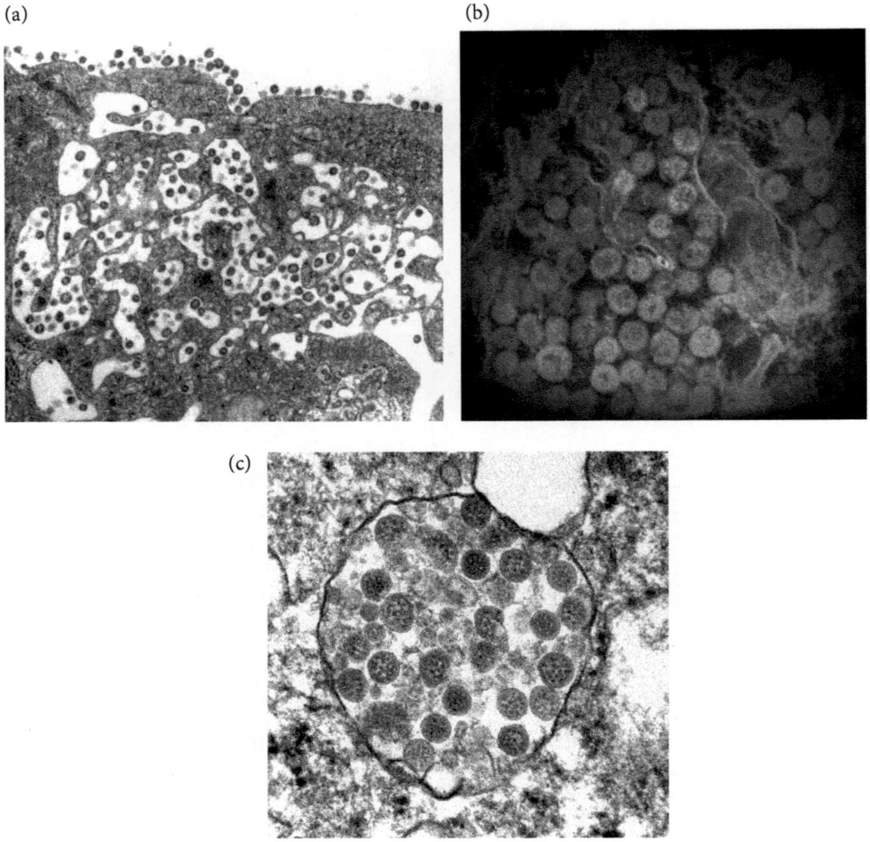

Figure 8.3 Coronaviruses multiplying inside a host cell. (a) SARS (2002–3); (b) MERS (2012–13); (c) Covid-19 (2019–ongoing)

probably look back on these images and say, "People used to think these were coronavirus virions!".' This explains why this book doesn't need a theory or definition of 'observation': it is uncontroversial to call *the examples I make use of* examples of observation. And there are even uncontroversial examples of observation at a much smaller scale, the scale of DNA itself (Figure 8.7).

This gives a snapshot of what scientists are drawing on when they analyse a new case. Their observations thus reveal things *to them* that would be absolutely opaque to a layperson. Combining the symptoms of the patients including how they develop over time, the indirect techniques probing both virus characteristics and immune response, and robust observation of the virus in/around precisely the cells expected given the symptoms, scientists can establish a basic causal relationship between the virus and the illness very quickly

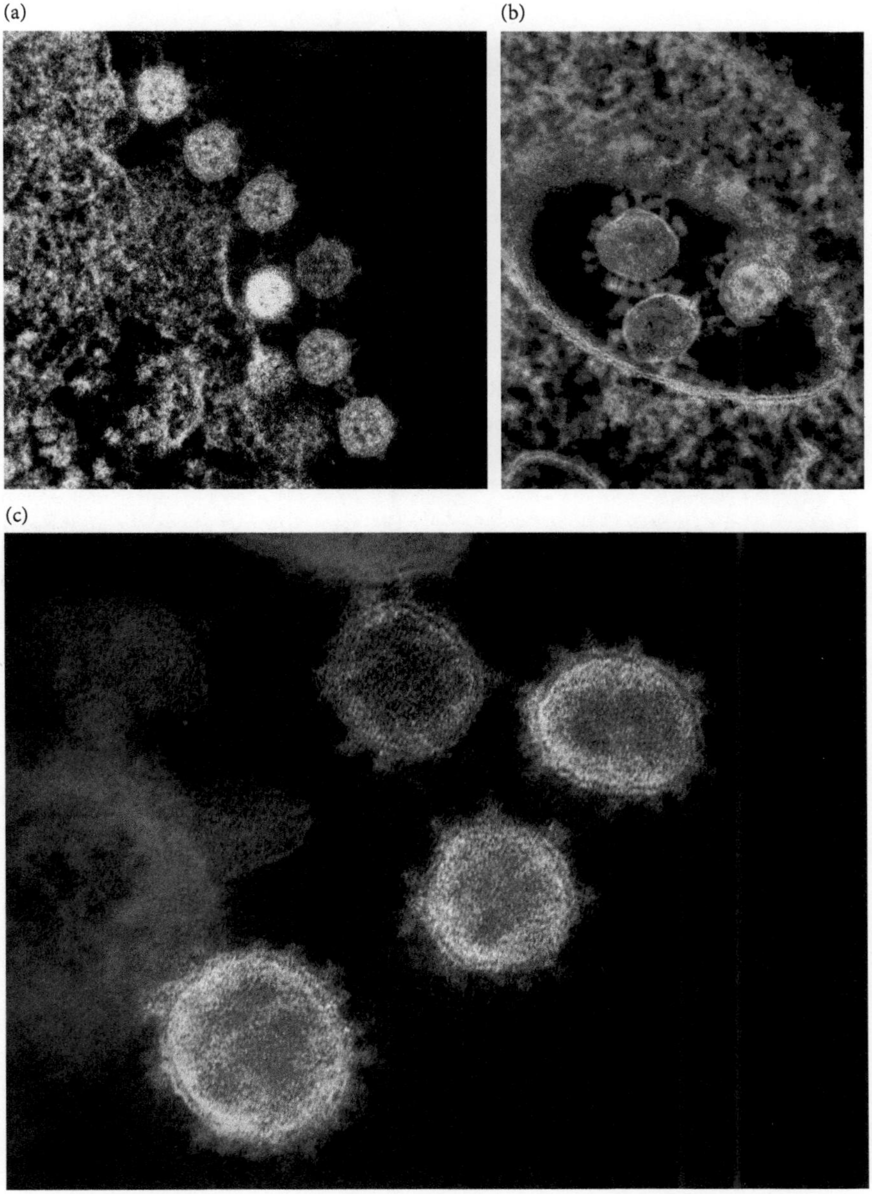

Figure 8.4 Three different types of coronavirus virion, all of which have demonstrated the capacity to infect humans. (a) SARS-CoV; (b) MERS-CoV; (c) SARS-CoV-2

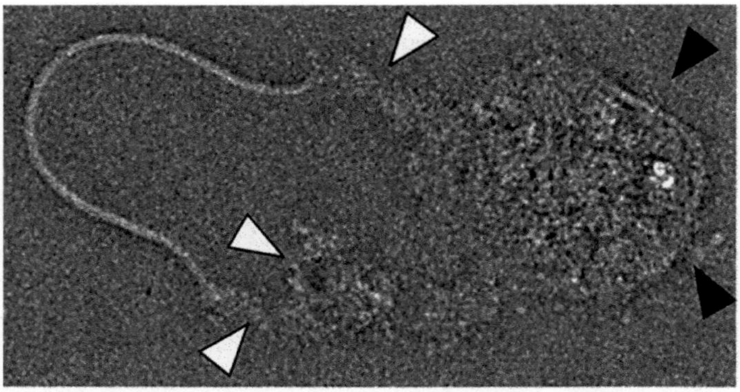

Figure 8.5 A SARS-CoV virion (RHS) ejecting its genetic material (LHS); for present purposes the arrows can be ignored

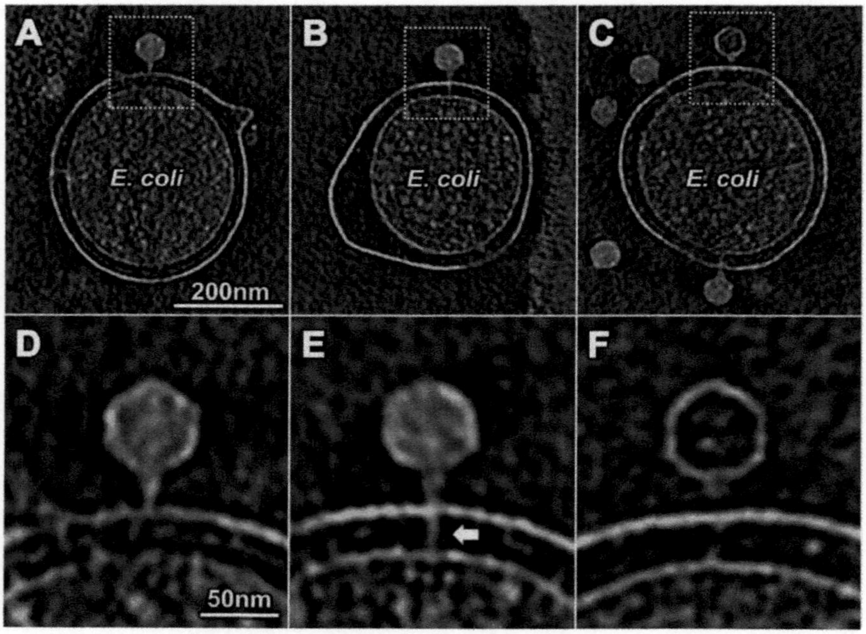

Figure 8.6 Real images of a virus particle infecting a cell, in three stages: (i) A&D, (ii) B&E, (iii) C&F; the arrow shows the moment infection occurs

indeed. *Precisely* what it means to say that the virus causes the illness can remain unclear. In the case of SARS, for example, in most fatal cases, the patient died some days or weeks *after the virus had been defeated* (Nicholls et al. 2006). It can be unclear what is most to blame: the virus, or the immune response. Thus the full causal story might be elusive, even some

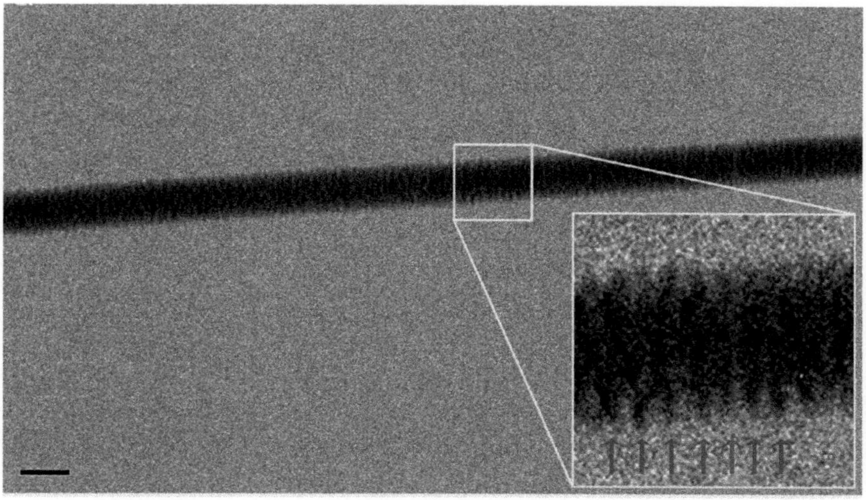

Figure 8.7 Transmission electron microscopy partially reveals the double-helix structure in a strand of DNA

years after the *basic* causal claim regarding the virus and the illness has been definitely established.

3. The Mesosome Objection

I've put quite a lot of emphasis on observation in this chapter, and especially observation via electron microscopy. However, observation by use of an electron microscope is very far removed from observation with the naked eye, and there have been problems in the past, lessons to learn from. A particularly famous example concerns the story of mesosomes. The features of cells known as 'mesosomes' were observed with electron microscopes in the 1950s, 1960s, and 1970s, and they were taken very seriously. The idea that they are *non-artefactual* real features of cells was later abandoned, however. Rasmussen (1993, p. 227) goes as far as to say that the mesosome 'ended up an artifact after some fifteen years as a fact'. Thus here we meet with a challenge to my appeal to 'observation' as a short-cut to future-proof science. The mesosome saga apparently shows that what we see with the electron microscope might be misleading: it might be something that *we* introduce to the target of observation, without realising it. If Rasmussen is right, then even something secure enough to deserve the word 'fact' might be overturned, many years down the line. This is a potentially serious challenge to my appeal to 'observation' to support claims about SARS-CoV-2 as the cause of Covid-19: one can imagine a sceptic using the mesosome case to try to cast doubt on claims about SARS-CoV-2. It is also a potentially serious challenge to 'future-proof science' generally: for Rasmussen, scientific 'facts' come and go, and the mesosome case is just one example of many. One is here reminded of the Oreskes quote discussed in previous chapters: 'scientific truths are perishable' (Oreskes 2019, p. 50). This is an alarming point of agreement between two scholars—Rasmussen and Oreskes—who are diametrically opposed in many ways.[7]

There have been responses to Rasmussen's claims in the literature, most obviously Culp (1994) and Hudson (1999, 2013). But neither author objects to Rasmussen's claim that the existence of mesosomes was a 'fact'. Instead, the focus of Culp and Hudson is on the scientific process in the 1970s that served to establish that mesosomes are artefacts; they wish to respond to Rasmussen's social constructivist claims that, in the mid-1970s, '[T]he scale could have tipped either way. Mesosomes could be accepted as real' (p. 256) and '[T]he scientific truth is nothing more than what the community of scientists ends up agreeing upon' (p. 265). On the question of how securely the reality of mesosomes was established in the 1960s and early 1970s, Hudson writes as follows (in apparent full agreement with Rasmussen):

> Pictures of mesosomes were produced by electron microscopists from the 1950s through to the mid-1970s, with hundreds of papers appearing in prestigious journals containing experimental results describing mesosomic structure, function and biochemistry. After 1975, however, the views of the microbiological community changed: Mesosomes were no longer asserted to be bacterial organelles but rather claimed to be artifacts of the process by which bacteria are prepared for electron-microscopic investigation.
>
> (2013, p. 55)

Rasmussen's original paper (1993) details a considerable fraction of the scientific activity, as it appears in the journals. He then writes, 'In this large body of work extending into the '70s, of which the cited studies are only a representative sample, the topic at issue was the biological function of the mesosome' (p. 246). As for its bare *existence*, that was a 'fact', and wasn't even

[7] It is no coincidence that Rasmussen and Oreskes are both *historians* of science. The claim that the history of science shows that 'the future always tends to outwit scientists' ephemeral certitudes' is an 'all-too-common historiographic trope' (Engstrom 2016, p. 179).

debated, according to Rasmussen. And he certainly provides some evidence to support his claims. For example, continuing on p. 246 he references a review article by Rogers (1970), published in *Bacteriological Reviews*, containing the following: 'Most of what has been said concerns only gram-positive cells. In gram-negative species, mesosomes also undoubtedly exist' (p. 202). The key word here is, of course, 'undoubtedly'. And Rasmussen calls this the 'standard opinion' (p. 202).

Of course, as already discussed at length in this book, it isn't enough to find one or two quotes in the literature. The opinion of one or two individuals is completely negligible compared with the entire community of relevant scientists, which in this case certainly consisted of many thousands of researchers. Thus a key move for Rasmussen is the step from the fact that many scientists were researching the *function* of mesosomes, to the claim that these researchers didn't doubt their *non-artefactual* existence. He writes, for example, that, 'Work on the entity involved the best efforts of dozens of important microbiology research groups around the world' (p. 227) and, 'All [of the biochemists involved] were de facto supporters of the existence of the mesosome by their choice to study the entity's composition, in much the same way as the biologists who were exploring mesosome function' (p. 250).

But this is a controversial move. Just because many relevant scientists were researching the function of mesosomes doesn't mean that they considered their *non-artefactual* existence an established fact. Mesosomes really did exist: the preparation process introduced the structures known as 'mesosomes' into the cells. It's not as if they were illusions, or tricks of the light. Anyone who said, at the time, 'Mesosomes are real!' was in fact correct—they *were* real, but they were artefacts, created during the preparation process. Rasmussen makes no attempt to emphasise this crucial point.

In addition, during the period in question, especially the 1960s, the 'mesosome' was a target of interest, to be explored by a range of new techniques that had just been developed, or were being developed, at the time. It was a classic period of what has been called 'exploratory experimentation' in literature such as Steinle (1997, 2002), Burian (1997), Franklin (2005), Elliott (2007), Waters (2007), and Karaca (2017). Without going into this literature in detail, the thrust of exploratory experimentation is nicely summarised by Franklin (2005): experiments can sometimes take on the role of 'exploring new phenomena when theories are either absent or in turmoil' (p. 888). This is certainly true of the kind of mesosome research in the 1960s so thoroughly referenced in Rasmussen (1993).

Indications that we are here dealing with 'exploratory experimentation' are numerous. The microscopy techniques were new, and were being constantly adjusted and developed, something Rasmussen (2001, pp. 633, 641) acknowledges. Relevant practitioners obviously knew this, and sometimes they explicitly commented on how tentative any claims about the ultrastructure of cells, including mesosomes, were. Even Ryter—one of the co-developers of the Ryter–Kellenberger (R–K) method that was widely used to investigate mesosomes—was cautious about the status of any claims. In a review article in 1969, she wrote as follows:

> It must, however, be pointed out that the behaviour of the mesosomes and their relation with the nuclei must be interpreted with great caution. [...] [I]t is obvious that the interpretations of morphological observations made till now can be considered only as working hypotheses.
>
> (Ryter 1969, p. 172)

At the very end of the review, she writes, 'For the time being they [mesosomes] cannot be considered as true organelles having a specific function.' And throughout the review there are numerous turns of phrase indicative of exploratory experimentation, including 'These conclusions are still preliminary' (p. 165), 'The process of nuclear division... is still poorly understood' (p. 170), and 'these actual data and knowledge about the respiratory chain [respiration of the bacterium] are too scanty and fragmentary for any definitive assumption' (p. 172).

Many others were of a similar opinion, including many of those, referenced by Rasmussen (1993), who were heavily involved in investigating mesosomes *as if* they were real, non-artefactual structures. Those involved knew that they were dealing with 'exploratory experimentation', even if they didn't use that term. In 1962, Van Iterson—another author Rasmussen references in support of his conclusions—wrote, 'It is now becoming most fascinating to consider the various cell constituents of micro-organisms... but we are still very much in the beginning' (Van Iterson 1963). To be blunt—contra Rasmussen (1993)—scientists are not inclined to speak of established 'facts' when they consider their research programme to be 'still very much in the beginning'. Van Iterson's paper was in fact presented as part of a symposium entitled 'Membrane Permeation' at the VIII International Congress of Microbiology, held in Montreal, Canada, in 1962. To introduce the session, the Chairman stated clearly that 'little is known of the mechanisms involved', and 'The techniques of electron microscopy have developed rapidly.' A lack of knowledge, combined with rapid development of techniques, is prime territory for the kind of 'exploratory experimentation' that does *not* go hand in hand with confident claims concerning future-proof scientific 'facts'.

Rasmussen's references and quotes from the literature are always in service of supporting his claim that the non-artefactual existence of mesosomes was a 'fact'. This claim—that the existence of mesosomes was a 'fact'—is an absolutely necessary premise of Rasmussen's wider ('social constructivist') agenda, and this is reflected in the presence of the word 'fact' in the title of his paper. But it is possible to provide an alternative perspective on the relevant microbiology literature, suggesting that there was plenty of caution in the community, and that far more than 5 per cent of the community would *never* have said that the non-artefactual reality of mesosomes was beyond doubt. In addition to the quotes and references supplied above, one can add the following examples.

Steed and Murray (1966) is published right in the middle of the period when Rasmussen claims that the non-artefactual existence of mesosomes was an established scientific 'fact'. But we find considerable caution in this paper, including explicit concerns about artefacts. In the abstract we find: 'attention is drawn to the problems of artefact in the preservation and study of dynamic structures at high resolution.' Perhaps the most important source of this worry is this: 'Light microscopy has not provided us with adequate controls to electron microscopic observations' (p. 264). As a result, 'one can do no more than guess at what is happening in the normal cell' (p. 264). This bears heavily on Rasmussen's central claim: Rasmussen would have us believe that there was no serious community doubt about the non-artefactual reality of mesosomes as 'organelles' of cells, at the very same time that Steed and Murray were writing that 'one can do no more than guess at what is happening in the normal cell.' Are we to believe that Steed and Murray were radical, 'fringe' sceptics?

To give another example, in Patch and Landman (1971) we find much caution and even plain confusion about what is going on in the field of investigation. They mention 'several tentative conclusions' (p. 355), note that 'There is no clearcut agreement between different laboratories' and say, 'It is not profitable to dwell on these differences, since the analytical data vary quite markedly depending on [several factors]' (p. 355). They continue to describe what they consider to be a 'chaotic picture' of 'inchoate data' (p. 355). They finally state: 'We conclude that the presently available data on the enzymatic activities or the labeling of mesosome vesicle proteins and lipids do not provide a substantial clue to the role of mesosomes' (p. 356).

One possible response Rasmussen might still try is to claim that most of this confusion and doubt merely pertained to the *role* of mesosomes in the cell, *not* to their bare existence. Now, just to emphasise once again, mesosomes *did* exist, so Rasmussen's story can*not* be that scientists thought that mesosomes existed, but now we know that they don't. But Rasmussen would no doubt insist that since researchers were asking the question 'What is the role of the mesosome in the cell?' they were thereby assuming that mesosomes existed in *all* cells, not only those that were prepared (e.g. using the R–K method) for microscopic observation.

But how much community confidence was there that the preparation techniques were not introducing artefacts? Many experiments of the day were explicitly devoted to the question of the effect of the preparation techniques on the cells to be observed. And the researchers involved often stated quite explicitly that much was still unknown concerning how the preparation techniques affected the cells. Steed and Murray (1966), focusing on cell division, wrote that, 'We have, therefore, been encouraged to [...] consider the possibility that *our present fixation procedures may be inadequate* for the preservation of such dynamic events' (emphasis added). In 1969, Burdett and Rogers were explicitly investigating the extent to which the procedures were affecting the very thing to be observed: the title of their paper was 'Modification of the Appearance of Mesosomes in Sections of *Bacillus licheniformis* according to the Fixation Procedures' (published as Burdett and Rogers 1970). Of course, one could try to claim that these authors were 100 per cent sure that mesosomes exist as non-artefactual features of cells, and they were merely investigating the extent to which the appearance of mesosomes is *modified* by the fixation procedures. No doubt some in the community thought that way. However, many others were explicitly concerned about the extent to which the preparation procedures were affecting the cells *tout court*, not merely the appearance of mesosomes.

If there was really doubt about the non-artefactual (*in vivo*) reality of mesosomes, it ought to be possible to find at least some practitioners stating that explicitly in print. Rasmussen (1993) wants us to believe that such doubts weren't expressed by anybody in the community until at least 1971. He writes:

> Doubts were first cast on the mesosome by microscopists exploring variations in the Ryter–Kellenberger fixation method. A 1971 paper by Silva found that changes in the buffer recipe, such as omission of calcium, caused variation in mesosome morphology. (p. 251)

And he adds, 'The first direct question (in print) as to whether mesosomes really exist was raised by Nanninga in 1973' (p. 251). The references are to Silva (1971) and Nanninga (1973), but it must be stressed that Silva and Nanninga were both casting doubt on the non-artefactual reality of mesosomes in the 1960s. Silva's earlier work—not referenced by Rasmussen—is particularly noteworthy. In a series of talks and papers going back at least as far as 1966 Silva repeatedly raised concerns about the effect of the preparation techniques on the target of observation. In Silva (1966a, 1966b) he was already quite explicit; the following (all taken from the Silva 1966a abstract) provide a flavour:

> [T]he influence of the various factors involved in fixing mesosomes is not well known.
>
> The morphology of the mesosomes of this bacteria [*Bacillus cereus*] is different depending on the fixation.
>
> These results suggest that the structure of *N. asteroides* mesosomes may change during fixation due to the conditions under which fixation is performed.
>
> In conclusion, the observations presented show that the bacterial membranes are sensitive to the conditions of fixation to a much greater degree than that found with the membranes of eukaryotic cells and lead us to admit the possibility of the aspects of mesosomes presented by several authors in bacteria fixed by some of the processes currently used do not correspond to the state of these structures in living cells.

Silva followed this line of research for several years, presenting his work to the community each and every year. In all cases the take-home message was that the fixation procedures were affecting the cells, and thus little, if anything, could be said with confidence concerning the microstructure of the cells observed, with a special focus on 'mesosomes'. In 1967, at the 2nd annual meeting of the Portuguese Society for Electron Microscopy (SPME), Silva was involved in three of the presented papers (Silva 1967a, 1967b; Silva and Guerra 1967). In each of these the focus was the same: exploration of the effect of the fixative procedures on the targets of observation, with explicit focus on mesosomes. This continued at the 3rd and 4th annual meetings of the SPME (Silva and Mota 1968; Silva et al. 1969), and at the 5th annual meeting of the SPME in 1970—in joint work with A Freitas Da Fonseca—he was explicit about the 'ultrastructural membrane damage' caused by the fixation methods (Silva and

Da Fonseca 1970). By 1971—in joint work with Dinah Abram—he wrote explicitly of 'artefacts':

> Taking together the results obtained by the variations in the conditions of chemical fixation with those observed with the different procedures of negative staining, the conclusion emerges that the possibility should be considered of the IMS [intracytoplasmic membrane systems] widely described among Gram positive bacteria prepared by the conventional procedures being artifacts induced by these methods.
>
> (Silva and Abram 1971, abstract)

In that same year he had success with a publication in one of the forefront journals in his field (Silva 1971), and explicitly advocated extreme caution concerning any claims about mesosomes:

> It is difficult to ascertain which, if any, of the membrane configurations obtained by the changes in the OsO_4 fixation, is real. [. . .] In conclusion, it is clear that, at present, no confident picture of bacterial membrane ultrastructure can be deduced from chemically fixed, negatively stained or freeze-etched preparations, since a consistent pattern is not observed. (p. 231)

Rasmussen would have us believe that before 1971—between 1965 and 1970, say—the non-artefactual existence of mesosomes was a 'scientific fact'. But it is crucial to recognise that some in the community were spending the 1960s merely conducting experiments to explore the possibilities. During this time of exploratory experimentation, it is perfectly normal to find a lack of explicit statements of confidence that mesosomes are artefactual; Silva is an exemplar of those in the community who were simply taking their time, making observations and conducting experiments.

One could tell a similar story about the trajectory of N Nanninga's research. The key quote in Nanninga (1973) is:

> At first sight it might seem surprising that the large structures described as mesosomes in osmium tetroxide fixed cells might be artifacts. However, it should be recognized that treatments of bacterial cells during and after collection are extremely long with respect to the life span of a bacterium. Under such circumstances physiological adjustments eventually followed by degenerative effects are not at all unlikely.

This quote expresses a serious degree of doubt regarding the (non-artefactual) reality of mesosomes, but note that such a serious degree of doubt is not required to cast sufficient doubt on the (non-artefactual) reality of mesosomes such that we should hold back from using the word 'fact' when we consider their non-artefactual existence in cells. And anyway, in 1971 Nanninga had already expressed very considerable doubt, writing:

> The reliability of structures observed with electron microscope techniques represents a general problem in ultrastructure research. Obviously, every technique employed will have its limitations. Unfortunately the limitations are incompletely understood, and the eventual image obtained by electron microscopy deviates to an unknown extent from reality.
>
> (Nanninga 1971, p. 219)

This was published in 1971, but concerns earlier work. Rasmussen overlooks this article.

Thus Rasmussen is far too keen to find a scientific fact here. It is trivial that, if one sets the bar low enough for a 'scientific fact', then one will find many 'facts' that have subsequently been overturned in the history of science.[8] As stressed repeatedly in this book, I am setting the bar very high indeed. As a result, I may agree with Rasmussen completely that, in the 1960s, the majority of relevant scientists fully believed in the non-artefactual existence of mesosomes. But the *mere* majority might only be 51 per cent of the community! The crucial criterion for future-proof science put forward in this book is that 95 per cent of the relevant community would be prepared to state that the claim in question is a 'fact', and that debate on the matter should be closed down. Perhaps Rasmussen would fully agree that the high bar I set in this book was *not* reached for mesosomes. In his 2001 paper responding to Culp and Hudson, he writes, 'In 1960s theories that membraneous organelles such as mesosomes are real were better supported than at other times' (Rasmussen 2001, p. 639). I absolutely don't doubt that, but this is very poor support for use of the word 'fact'. He also writes, '[F]or a long time most [microbiologists] did [believe in mesosomes]' (p. 646f.). But again, 'most' is nowhere near sufficient for future-proof science. The slide from 'Most scientists think X' to 'Scientists think X' is a serious error we have come across a few times in this

[8] Cf. Tucker (2003) responding to Solomon's (2001) list of examples of past scientific consensuses that have been overturned: 'With a vague concept of consensus, it is possible to adapt the judgement of what is considered a case of consensus or dissent to the argument one wishes to make' (p. 509f.).

book already. Whilst the former is consistent with a mere 51 per cent of scientists favouring X, the latter suggests a consensus. The difference for epistemology is enormous.

Summing up, this section presents reasons why the *non-artefactual* existence of mesosomes should never have been considered future-proof science. The case doesn't meet the criteria put forward in this book, primarily because it was a rather tentative period of exploratory science (even if *some* relevant scientists made rather bold statements at the time). But couldn't research into Covid-19 and SARS-CoV-2 *also* be considered a period of exploratory science? Weren't researchers faced with a significant novelty—a new illness and a new virus—and weren't they initially *exploring* the illness and virus?

A useful distinction can be made here between exploratory science and what Kuhn described as *normal* science.[9] Normal science can be characterised as a form of everyday scientific puzzle-solving using a bunch of tools (concepts, methods) that are accepted and shared within the community. In the case of Covid-19, the 'puzzle' was to understand the nature of the illness and its cause. This was thoroughly 'normal' science to the experts involved, since although the specific (token) illness was new, it fitted very neatly into a *type* of illness that had already been thoroughly researched for decades, namely, lower respiratory tract infections. Moving on to the cause, a novel coronavirus later named SARS-CoV-2, the investigation of this virus was also 'normal' science. Just as we can straight-forwardly fit the specific illness Covid-19 into a wider *kind* of illness, we can also fit the specific virus SARS-CoV-2 into a wider kind of virus, namely, coronavirus. As already detailed earlier in this chapter, coronaviruses had already been researched for decades. Thus we are not dealing here with what would properly be called 'exploratory science'. Recalling the Franklin (2005) quote mentioned earlier, exploratory science has to do with 'exploring new phenomena when theories are either absent or in turmoil' (p. 888). But the theory surrounding Covid-19 and SARS-CoV-2 was solid.

The mesosome case was completely different in the relevant respects. Mesosomes were a completely novel phenomenon that could *not* be placed within a well-established kind. Indeed, they couldn't be placed within *any* known kind. And any theories put forward about them were incredibly tentative, as the researchers sometimes stated explicitly in their articles. Even the Fitz-James (1960) paper Rasmussen puts so much weight on is thoroughly

[9] 'Normal science' is usually contrasted with 'Extraordinary science', but there needn't be anything particularly extraordinary about exploratory science.

speculative.[10] The puzzle of mesosomes was a novel puzzle for researchers, and the concepts and methods used to access them were in flux, most obviously because the microscopy techniques involved were in a process of rapid development. In short, there were numerous reasons for epistemic caution in the mesosome case that do not apply to the Covid-19 case, and there were numerous reasons for epistemic *confidence* in the Covid-19 case that do not apply to the mesosome case.

4. Concluding Thoughts

Naturally there are a huge number of questions one may ask about Covid-19, its cause, and how best to tackle it. Each and every query must be tackled separately, and fine distinctions between questions are important. For example, the question of whether Covid is caused by a virus is rather different from the question of whether Covid is caused by a coronavirus, which is itself rather different from the question whether it is caused by a *novel* coronavirus. In this chapter I hope to have merely given a sense of why the basic, broad-brush story of the viral cause of Covid-19 was future-proof a mere matter of a few months after the first cases of Covid-19 were announced in December 2019. I say a few months, because even in a case of 'normal' science as easy as this—a known type of disease along with a known type of virus—a period of peer scrutiny is still required. Crucially, what one will *not* find in the case of the cause of Covid-19, is a serious journal article *response paper* stating opposition to those early papers (e.g. Tan et al. 2020; Huang, C et al. 2020) that confidently identified the cause of the new illness as a novel coronavirus.

If I'm right that the best way to identify future-proof science is via scientific consensus, then the predominant strategy in use today when assessing scientific claims is problematic. The predominant strategy is to look to a *top expert*, and trust in their judgement. In the case of the Covid epidemic in the UK in 2020, the 'top expert' constantly appearing on our television screens was Professor Chris Whitty, Chief Medical Officer and Department of Health and Social Care Chief Scientific Adviser. Whilst there is no doubt that Professor Whitty is a good source of information, we've seen a few times in this book how trust in individual 'top experts' can be a poor strategy compared with an assessment of the *community* of relevant experts. One way this

[10] See, for example, the section entitled 'The Possible Functions of Mesosomes' (Fitz-James 1960, p. 522).

manifested itself in the case of the Covid pandemic was in the considered opinion of Michael Levitt, a Nobel laureate at Stanford University, and Fellow of the Royal Society. In March 2020 Levitt predicted that recovery from the pandemic in Italy and the USA would be quick and there was little to worry about. But Levitt, however respected, was just one individual. And in addition, as Allchin (2020, p. 2) relates, 'His Nobel Prize was in chemistry'. He was not known for modelling pandemics. Whitty is also just one individual, but at least his background is obviously in the right field. There is still a possibility that he will report to the public *his* view of the science, and this might not always line up with the *community* view. This is especially possible when there is a majority in favour of X (for any given X); in such circumstances there is that eternal tendency to slide from 'most scientists think X' to 'scientists think X'.

What *should* the layperson do, then, when assessing the state of play vis-à-vis some particular claim 'X'? My advice is to hunt down any serious scientific *resistance* to X. In the case of the Schulte et al. (2010) paper strongly favouring the Alvarez hypothesis, it doesn't take much hunting to identify the serious scientific community resistance to the claim (e.g. those three letters in *Nature*). In the case of the cause of Covid-19, even a *lot* of hunting around would *not* reveal any serious scientific resistance to the standard story, that SARS-CoV-2 is the cause. There are, of course, many other Xs, and every case will be different, but my suggested strategy remains the same: look for the serious scientific community resistance. This means avoiding articles written by journalists, except to use them to identify better sources, written by scientists. An example of a good source is *The Conversation*, where the articles are written by relevant experts, are accessible for non-experts, and include links to relevant scientific literature. In addition, one should always be on the lookout for an individual *merely asserting their individual opinion* (or the opinion of their particular research group), as opposed to an individual reporting overall community opinion.

In the case of the Covid-19 pandemic things have moved very quickly indeed. It's hard to keep up. This includes scientific community opinion on various issues, including the effectiveness of certain kinds of face mask, the vulnerability of various age groups, the emergence of 'new strains', and so on. In many cases, action needs to be taken *right now*, and so all one can do is trust the best evidence available at the time. We are mostly *not* dealing with science that is known to be future-proof. In fact, there *is* a significant amount of exploratory science taking place, especially when it comes to social science—for example, when we ask to what extent a society will tolerate different types of restrictions and lockdowns. But however much uncertainty we are faced with, it's crucial to establish rock solid foundations for debate if at all possible. In the case of the coronavirus pandemic we can establish some rock solid foundations: we really do know a lot about Covid-19, and the virus that is responsible. A large part of the reason that we know so much is because we had two very serious practise runs, with SARS in 2002–3, and MERS in 2012–13. It didn't have to be that way: the Covid-19 pandemic might have hit us *without any practise runs*. This stroke of luck for humankind is often overlooked.

9

Core Argument, Objections, Replies, and Outlook

1. Can We Identify Future-Proof Science?

At times this book has somewhat conflated two different issues: (i) that of establishing criteria that, when met, are sufficient for future-proof science, and (ii) the issue of how, in practice, to identify future-proof science. These can easily come apart. For example, some critics may say that *even if* the criteria I have put forward are indeed sufficient for future-proof science, it is often practically impossible to *ascertain* whether or not the criteria have been met. They might even hold up one of my historical examples (continental fixity? mesosomes?) and say that *if* we had been living through it (as opposed to looking back with the benefit of hindsight), we would almost certainly have thought that the criteria had been met. Says this critic: only in hindsight, knowing that the claims in question fell down, do we hunt around for evidence that one of the criteria wasn't met.

This final chapter starts by clearly separating these issues. I first (Section 1.1) present my final considered view on the criteria for future-proof science, and (Section 1.2) summarise the core argument supporting these criteria. I then move on (Section 1.3) to the question of how to work out whether or not the criteria are met, in practice. In Section 2, I turn to outstanding objections that have not already been addressed in the preceding chapters. Section 3 turns briefly to some important implications for interventions on science education in our schools. Section 4 considers future directions.

1.1 The Criteria for Future-Proof Science

A primary aim of this book is to present a filter, where any scientific claims that get through are certainly future-proof. And this isn't because we are stuck in a rut of human thinking, but because we have actually hit upon the truth, or

Identifying Future-Proof Science. Peter Vickers, Oxford University Press. © Peter Vickers 2023.
DOI: 10.1093/oso/9780192862730.003.0009

'facts'. Nor should 'truth' and 'fact' be interpreted in a *thin* way, compatible with Oreskes' (2019) claim that 'scientific truths are perishable', or Rasmussen's (1993) social-constructivist claims about 'scientific facts', suggesting that they come and go. The claim, instead, is that we are as certain about the status of those scientific claims as we are about historical claims such as the claim that the Second World War really happened, the Hiroshima bomb really was dropped on 6 August 1945, and so on. Quite obviously, it will *always* be accepted that the Second World War really happened, and the Hiroshima bomb really was dropped (assuming that the human race avoids apocalypse). Similarly, it will always be accepted that the Sun is a star, and viruses multiply inside host cells. The question is not whether there are *any* examples of future-proof science; there are a great many, and many of them are very easy to identify.

It is crucial to bear in mind that getting through the filter is sufficient, but not necessary, for future-proof science. After all, many scientific claims are currently *well on their way* to becoming established facts, but are not there yet (a stage in the process of becoming a fact that all of my 30 examples in Chapter 1 went through). We can't identify the specific claims right now, of course, but one good candidate is the Alvarez hypothesis, which currently enjoys a great deal of support within the relevant scientific community (as discussed in Chapter 7). It doesn't yet get through the filter, but that doesn't mean it isn't a future-proof claim; we don't know either way right now. It is also crucial to bear in mind that choosing the 95 per cent threshold is *not* inconsistent with believing that, really, we should be sure to the extent that the consensus is strong.[1] I think that's right, but in this book the claim is merely that what makes it through the filter—complete with 95 per cent threshold—is in fact future-proof.

The two criteria for future-proof science are more complex than they initially appear. Simply put, they are:

(1) 95 per cent of the relevant scientific community must be willing to describe the claim in question as an 'established scientific fact'.
(2) The relevant scientific community must be large and incorporate a substantial diversity of perspectives.

[1] Cf. Miller (2013), p. 1313f.: '[T]he conditions for knowledge-based consensus are a matter of degree. The more they are met, the more likely a consensus is knowledge based.'

But there are significant complexities once one starts unpacking these two criteria. There are some potentially tricky key words, concepts that non-trivial weight is put upon. One is 'relevant scientific community'; a sceptic might allege that until this is unpacked my thesis cannot be properly assessed. But I resist this: in any specific case there are obviously relevant scientists, and obviously irrelevant scientists. When it comes to borderline scientists, there are reasonable and unreasonable places to draw the line. Thus a detailed philosophical analysis is not required. If a case is put forward that challenges the proffered link between the criteria and future-proof science, then in the fullness of time it will be obvious if the only way to wriggle out of the challenge is to draw the 'relevant scientist' line somewhere that is widely considered *un*reasonable. And I would offer similar responses to calls to unpack other key terms.[2]

In the course of writing this book I considered quite a variety of additional criteria. The reader may be wondering about other such possible criteria, too. Here are some of the front-runners that didn't make the final cut:

- *The consensus of opinion shouldn't be mere agreement, but must be based on properly scientific activity.* This is true, but it isn't needed as an additional criterion, since criterion (1) demands that the claim is considered an 'established scientific fact', ruling out cases where the statement in question is a mere shared assumption (say). This covers cases such as the widely shared (false) assumption *c.*1990 that the expansion of the universe is *not* accelerating. It was never considered an *established scientific fact* that the universe's expansion is not accelerating; far from it.
- *The claim must have stood the test of time; at least a few years of international scrutiny must have taken place.* First of all, the latter clause is unwarranted (as discussed in Chapter 8). The first clause is important, but completely vague, and thus unhelpful as a criterion. How long should we wait? This is highly contextual, and the relevant community will decide. Once criterion (1) is satisfied, enough scrutiny has taken place, and enough time has passed. Thus criterion (1) does the work here.
- *The science in question mustn't be 'exploratory science'.* But this isn't needed, since the relevant community knows when it is involved in *exploration*, and claims are accordingly sufficiently tentative (or most of

[2] The unpacking of key terms is still important, though. Some literature that has attempted to do just this is mentioned in Section 4, below, where I consider possible extensions of my project.

them are), so that any consensus will fall short of 95 per cent (as seen with the case of mesosomes in Chapter 8).

- *The claim in question must be relatively modern (e.g. not prior to The Scientific Revolution).* This isn't needed, since if we go back in time too far either (i) it couldn't reasonably be called an 'established scientific fact' (e.g. the 16th-century assumption that the Earth doesn't turn wasn't an 'established scientific fact'), or (ii) there was an obvious lack of diversity in the community (e.g. the community wasn't remotely international, or was made up of 99 per cent men, etc.).

Thus I submit that the two stated criteria (1) and (2) are by themselves sufficient for future-proof science.[3]

Some scholars will be keen to ask what exactly it *is* about the noted kind of consensus that guarantees future-proof science. As discussed in Section 4 of Chapter 5, it is a reasonable tactic to distance oneself from this question for now. There is very significant precedent for such a tactic. It was wise in the decades following the publication of Newton's *Principia* to emphasise that massive bodies *just do* attract each other, even if we have no idea *why* they would do that. It was wise in the decades following the publication of Darwin's *Origin of Species* to emphasise that organisms *just do* evolve, even if we can say little or nothing concerning the underlying mechanism(s). And it was wise in the 1950s (say) to emphasise the evidence that the continents *just are* 'drifting', even if we have no idea how that could even be possible. Sometimes the deeper questions can wait.

1.2 The Core Argument Behind the Criteria

It is claimed that the criteria are sufficient for future-proof science. But what is the core argument supporting this claim? In truth there are several lines of argument, so a succinct 'core argument' shouldn't be demanded. However, it is perhaps worth here describing some of the most important considerations.

A key part of the argument is based on the list of 30 examples in Chapter 1. And yet we are certainly *not* dealing with a crude inductive inference from those examples to future examples, comparable with seeing 30 white swans

[3] However, there is no need for me to be deeply committed to never needing to add a third criterion. And future scholarship may demand adjustments to criteria (1) and (2). The core idea of this book wouldn't be significantly harmed by such developments.

and inferring that all swans are white. What's crucial here is that the criteria for nearly all of my 30 examples were met a very long time ago, and since then there has been enormous opportunity for the claims in question to be overturned. The exponential growth of science is relevant here, since that adds greatly to the volume of scrutiny these claims have withstood: roughly speaking, in most cases, one should compare the scientific work undertaken in the past 20 years with that undertaken in the previous *40* years. And in most cases the evidence base that was sufficient for a scientific consensus to form at time *t* has *more than doubled* in the years since time *t*.

If that were all we had to go on, some would still consider the argument to remain weak. They might draw on the work of Stanford (2019) and others who have emphasised just how conservative modern science can be, and how difficult it is to 'break the mould' and seriously pursue possibilities at odds with accepted wisdom. According to this kind of sceptic, a huge amount of scientific work does *not* mean that the claims in question have been seriously tested: since the claims are fully accepted by the community, hardly anybody is looking at them critically anymore. Herzenberg (2008), drawing on Gribbin (1995), writes, '[S]cientists can be about as gullible as other people in accepting ideas simply because "everybody knows" that they are true' (p. 17). Anyone who *does* dare to look critically at a 'fact' is either attacked or (more likely) ignored by the vast majority, who consider the debate closed down.[4] Thus it is no surprise that the claims in question haven't been overturned, *however* much scientific work has been undertaken in the relevant topic area.

But at this point in the dialectic we must ask ourselves just what *degree* of scepticism we need to overcome. For example, there is no need to lock horns with the 'radical' or 'global' sceptic (as discussed in Chapter 1). Nor should we expect to convince a dedicated flat-earther, or similar kind of scientific sceptic or conspiracy theorist. Instead, the intention is to respond to the position—out there in the literature—which presents itself as a 'rational', or 'reasonable' brand of scientific scepticism, championed by figures such as Hesse (1976), Van Fraassen (1980), Laudan (1981), Stanford (2006), Wray (2018), and Rowbottom (2019). What's crucial now is that most of these 'anti-realists' would *not* seriously doubt the truth of (most of) the 30 examples I presented back in Chapter 1, even if it often *sounds* as if they would (as discussed in detail in Chapter 2). And with the truth of those examples accepted, one *can* use

[4] Another interesting passage on this theme is found in Bullard (1975), who writes, 'it is more prudent to keep quiet, to be a moderate defender of orthodoxy, or to maintain that all is doubtful, sit on the fence, and wait in statesmanlike ambiguity for more data' (p. 5). Either way, the bold 'maverick' goes unsupported.

them to make the link between my criteria and future-proof science. (One remaining reason for scepticism vis-à-vis some of my examples will be tackled below, in Section 2.4.)

For the more radical sceptic, who does harbour serious doubts about several/many of the 'scientific facts' in my list of 30 examples, a further step can be taken. A crucial consideration comes from the fact that since the consensuses associated with those examples formed, new techniques have sometimes been able to essentially *prove* that the claim is true, via (sufficiently) direct observation. In Section 4 of Chapter 5, I mentioned nine such examples, where what was once theoretical has now been observed. Of course, these observations were not significant epistemic events, since by the time they came the relevant scientific community was already certain. As McMullin (1984) already wrote of this process many years ago, sometimes 'theoretical entities previously unobserved, or in some cases even thought to be unobservable, are in fact observed and the expectations of theory are borne out, to no one's surprise' (p. 33). But what this process does show is that the certainty in the community was justified, and rather than merely being a psychological fact about the community, a correspondence really does exist linking up the relevant belief with features of the world. This 'proof by observation' thus provides an extra reason—for anyone who needed it—to believe that meeting my criteria is sufficient for future-proof science. The nine examples in question were significant tests of my sufficiency claim, and no counterexample emerged.

1.3 Identifying Future-Proof Science in Practice

The primary aim of this book has been to provide criteria for future-proof scientific claims. But it is one thing to provide such criteria, and quite another to provide practical guidance on how to identify when the criteria have been met. If I ask myself the question, how would my own parents ascertain whether some given scientific claim 'X' they care about (e.g. concerning Covid-19) is a fact, I feel unsure. I could give them my criteria, and tell them that no claim has ever met those criteria and then later been overturned, despite truly enormous opportunity for such overturning, were it ever going to happen. But then they would be faced with the challenge of trying to ascertain how some huge, highly complex scientific community feels about claim 'X'. Would 95 per cent of that community be willing to describe 'X' as an established scientific fact, or not?

There are of course easy and hard cases. In some cases—such as Covid-19—scientific opinion is changing rapidly, and the burgeoning literature may be somewhat impenetrable to anyone outside of the relevant community. And in fact, when the scientific field is in flux in that way, even senior scientists in the field may genuinely not know whether or not 95 per cent of relevant scientists would assent to some particular claim. Serious opinions get expressed at conferences and in the literature, primarily, and it may take some time to establish whether or not we should be cautious about a particular claim that is being made. So in some cases all we can really do is be patient, and wait to see what happens in the community. If at least 5 per cent of the community dispute a stated claim, that will emerge eventually, in the form of formal debate at conferences and in the literature.

Will serious community debate be identifiable for the outsider? We saw in Chapter 7 how one could go about identifying the ongoing debate concerning the cause of the dinosaur extinction. There are different possible avenues, in fact. For example, one could look at the relevant major conferences, and find out what topics were presented over the past several years. Chances are, if there is debate about a particular claim, it will have been discussed at the relevant conferences. In the case of the dinosaur extinction this is quickly obvious when one looks at past conferences of the Geological Society of America (accessible on a smartphone with a simple search). Alternatively, one could follow the references on Wikipedia, and find the debate (if it exists) in the professional literature. In Chapter 7, this method was shown to be particularly robust vis-à-vis the dinosaur extinction debate. The Wikipedia page itself was somewhat misleading (when I last accessed it), since it suggested a strong consensus of opinion in favour of the Alvarez theory, citing Schulte et al. (2010). But it takes only a few seconds to click on the link to the Schulte paper, and find the three letters Archibald et al. (2010), Courtillot and Fluteau (2010), and Keller et al. (2010) that immediately opposed any suggestion of a solid consensus of opinion. So any flaw in the Wikipedia page is quickly circumvented.

One good rule of thumb when trying to ascertain whether opinion has reached 95 per cent is this: in most cases where it has *not*, evidence of substantial debate in the community will be relatively easy to find, and in most cases where it *has*, any serious opposition (within the relevant scientific community) will be extremely difficult to find. After all, in many cases the degree of opinion in favour of a claim will lie rather far from the 95 per cent cut-off, either far below, or close to 100 per cent. In relatively few cases will opinion lie quite *close* to 95 per cent. The Alvarez theory discussed in Chapter 7 is particularly interesting, since that *does* seem to be a fairly borderline case. In fact, whether or not 95 per cent has been reached in that case was *not* definitely established in Chapter 7. A *fairly* significant opposition to the Alvarez theory was identified, but I also presented some reasons to believe that, perhaps, a relatively *small* group of opposition scientists are managing to make it look like there is greater opposition than the reality. So everything I said in that chapter is consistent with 96 per cent of relevant scientists being ready to call the Alvarez hypothesis an 'established scientific fact'. But it's also consistent with a figure of only 94 per cent. Since there is currently some significant uncertainty, it is better to remain cautious: we can't allow the Alvarez claim through the future-proof filter just yet.

A crucial consideration in all this is to *find the relevant scientific community*, and look at what they are doing and saying. As shown in the Alvarez case, one can use Wikipedia, but *only* as a route to relevant scientific articles. Similarly, 'popular' articles can be highly misleading, although they might be another useful route to relevant scientists and their serious journal article publications. There are also useful in between articles, such as those sometimes found in *Scientific American* (e.g. Brusatte 2015), where a genuine member of the scientific community writes for a wider audience. It is crucial to check the author, however, since *Scientific American* authors are not always professional scientists.

There are many other things we could consider here, but ultimately this isn't a work in the sociology of science. A full discussion of all the ways that one might identify a 95 per cent scientific consensus (when one exists) and identify a greater-than-5 per cent opposition to a claim (when one exists), must wait for another day. For an entry to the literature on the social epistemology of consensus, see Miller (2019).

2. Objections and Replies

In this section I briefly tackle four of the most significant objections I have thus-far come across. My responses are not intended to close down the objections, but should instead be understood as the start of a conversation.

2.1 'Truth is not decided by a show of hands'

A popular saying going back decades (possibly centuries) is that 'Truth is not decided by a show of hands'. There are various variations on this theme. Here are just a few examples, taken from both scientific and philosophical literature:

> [S]cientific truth is not established by consensus.
> (Bergin and Strupp 1972, p. 145)

> Scientific truth is not decided by majority vote, of course (ask Galileo).
> (Sharp 2009, p. 206)

> Scientific truth is not determined by consensus. In science the majority can be, and often have been, wrong. There is a great deal of peer pressure in the science community, which can stifle objections to a popular theory.
> (Elsdon-Baker 2009, p. 215)

> When everyone believed that the world was flat, they were wrong. The truth is not decided by a show of hands. (Curnow 2012, Chapter 4)

This is a natural reaction, and one I encountered first-hand when sharing an early version of this book with scientists. Many scholars take the basic point to be obvious, and pithily evidenced with a platitude such as 'Everyone used to think the earth was flat. They were wrong', or, 'Everyone used to believe in geocentrism. They were wrong.'

The first point to make is that the saying is very seldom employed *carefully*. Consider the Elsdon-Baker quote: in the first sentence the emphasis is put on scientific *consensus*, whereas in the second sentence the emphasis is put on 'the majority'. But I have stressed several times in this book already that there is a wide gulf between these concepts, of huge moment for epistemology: whilst 'the majority' might mean only 51 per cent, the term 'scientific consensus' is best reserved for cases where the *vast* majority of scientists would agree. In this book I have used the term 'solid scientific consensus' to refer to cases where there is a strong and unambiguous agreement amongst at least 95 per cent of the relevant (international and diverse) scientific community. The flat earth case mentioned by Curnow (2012) is immediately ruled out, since there wasn't anything like an international scientific community back then (Curnow isn't referring to the *modern* flat earth conspiracy theory, of course). The same goes for the geocentrism case mentioned by Sharp (2009): 'science', in anything like the modern sense of that term, was born during The Scientific Revolution (often associated with dates such as 1543–1687, making reference to the highly influential contributions of Copernicus and Newton). Galileo wasn't fighting against a solid scientific consensus, even if he was fighting against *a* consensus of opinion (with prominent religious figures playing a major role).

In many contexts where a show of hands is sometimes *literally* used to decide something (e.g. Board of Studies meetings in the Department of Philosophy I call home), a *mere* majority is sometimes considered sufficient.

Taken in this way, this book certainly does *not* advocate a 'show of hands' approach to future-proof science; far from it. It is much better to use the term 'scientific consensus'. But even this isn't ideal, since many scholars readily conflate 'consensus' with 'majority'. Thus I have often opted for the term 'solid international scientific consensus', where this is shorthand for the two criteria that must be met to pass through the filter.

It may be objected that by 'truth is not decided by a show of hands' there is a more basic point, namely that truth is accessed via *evidence*, not community *opinion*. After all, there is nothing contradictory about a situation where 100 per cent of a community think of something as an 'established fact', and yet they are all completely deceived. In this case we would have the strongest consensus imaginable, and yet we would *not* have truth. Relevant quotes in the literature are not hard to find; for example,

> [A]t the end of the day, the only evidence for the truth of scientific theories is the evidence that scientists use, and the only positive arguments for scientific realism are the arguments that scientists make. (Lipton 2001, p. 353)

> The only sensible way to find out whether a scientific claim is true is by looking at the original scientific evidence previously offered by scientists. (Enfield 2008, p. 891)

My primary response to such claims was detailed in Chapters 4 and 5. It is impossible, in practice, to 'look at the original scientific evidence', since there is far too much of it, and it is far too easy to come away with a biased view of the overall evidence base. But we can *indirectly* access the entire evidence base, from a wide variety of different perspectives, by focusing our attention on features of the relevant scientific community. Thus this book fully agrees with the thought that the first-order evidence is all important. The issue is how to *get at* that evidence, in practice, and evaluate it in order to judge whether the claims in question are truly future-proof.

Scientists tend to find this shift of thinking—from the first-order scientific evidence to the second-order, consensus approach—a difficult pill to swallow. After all, all of their training, and all of their career, has been firmly focused on the first-order evidence. Thus when I shared an early version of this book with scientists—in particular to get feedback on Chapter 7—it was common to hear back some variation or another on 'The truth is not decided by a show of hands.' One scientist in particular—prominent in the dinosaur extinction debate—simply replied, 'the speed of light is not decided by raising hands or casting a poll.' For that scientist, the speed of light is (obviously!) decided by carrying out certain experiments, such as those detailed in Froome (1958) and Evenson et al. (1972). These experiments *reveal the speed of light*; any opinion poll will merely measure scientists' opinions.

However, the interferometry experiments detailed in Froome (1958) and Evenson et al. (1972) did *not* 'decide the speed of light'. It was 'decided' many decades previously. And when one looks back at the history, one certainly does *not* find one journal article, or one experiment, where the speed of light was established. Rather, the establishment of the speed of light was a long process, with many hundreds of individuals involved, and many hundreds of scientific publications, both experimental and theoretical. A period of speculation and reasonable doubt very gradually gave way to a period of certainty. The truth was, indeed, ultimately decided by scientific toil, but my question concerns when exactly it could be said that the evidence was sufficient to reasonably describe the speed of light as an established, future-proof scientific fact—*without* merely saying, trivially, 'when the evidence was sufficiently strong'. The evidence is always highly complex and multifaceted—even for something as simple as measuring the speed of light—and it doesn't come with any guidance on how evidence should determine credence. By contrast, it is *relatively* straightforward (though still challenging) to determine when a 95 per cent solid international consensus has formed within the relevant scientific community. And this book has argued at length that *that* marker is sufficient for us to have full confidence that we have hit upon the truth.

Scientists will still feel conflicted, no doubt. Consider the scientist working on the Alvarez hypothesis. Every step of her training and career encourages her to digest and assess the first-order evidence. Suppose, based on her scientific work over many years, she sides with the majority, thinking that the Alvarez hypothesis is now an established scientific fact. But then suppose she takes off her scientist hat, reads this book, and is convinced that one should *not* declare something an established fact until 95 per cent of the relevant community are in agreement. In this case she firmly believes the claim *when wearing her scientist hat*, and she recommends caution *when she takes that hat off*. But now it sounds like she is conflicted. Can she compartmentalise in this way? Certain on a Friday, but cautious on a Saturday? It sounds bizarre.

Suppose, convinced by this book, she gives up on the traditional, *scientific* way of deciding facts. She agrees that since we haven't reached 95 per cent agreement—let's say the agreement only stands at 94 per cent(!)—we should remain cautious. Suppose other scientists in the community decide to follow suit. In that case, the percentage starts to come down. If *everyone* in the community takes the same approach, we move from 94 per cent consensus to the point where 0 per cent of the community would be willing to declare the Alvarez hypothesis an established scientific fact! There also seems to be no way back *up* from 0 per cent to 95 per cent. The evidence will keep coming in, and let's say it keeps getting stronger and stronger. But that won't move the relevant scientists to start using the term 'established scientific fact' since, influenced by this book, they will only do that when it is clear that 95 per cent of the community are (already) willing to use that term. We seem to have a rather odd little paradox here.[5]

Whilst an interesting thought experiment, I think *in practice* there is nothing to worry about, and this is purely hypothetical. Scientists will *not* start linking the word 'fact' to a certain strength of scientific consensus. They will continue to assess the first-order evidence, and that's also exactly what we want them to do. Indeed, that's what we *need* them to do, in order for the consensus approach to work. There will be times when the surest way to the truth, or to avoid falsehood, will be to consider the strength of the consensus (since any attempt to digest the first-order evidence will be inevitably partial, biased, etc.). Since many scientists are very strongly motivated to find the surest way to the truth, and to avoid falsehood, they may actually *want* to take the consensus approach put forward in this book. That is why we must hide this book from the scientists.

2.2 When Is a Scientific Community Sufficiently Diverse for Future-Proof Science?

Imagine if the sole criterion for future-proof science were: 'A scientific claim is future-proof whenever the evidence is sufficiently strong.' This could never fail, since in any case of an overturned scientific claim the advocate could shrug her shoulders and say, 'The evidence wasn't sufficiently strong.' The problem with this toy criterion is the word 'sufficiently': it is vague, and thus allows wriggle room for the advocate whenever the sceptic tries to present her with a counterexample. Another way to make the same point is this: How would we ever *know* that the evidence was sufficiently strong? The vagueness

[5] This paradox, such as it is, appears to be a special case of a wider class of paradoxes that have already been identified and discussed in literature such as Elga (2010). (With thanks to Chris Cowie.)

of the formulation means that it becomes impossible to see, with any useful degree of clarity, which claims are future-proof and which aren't.

But now, turning to the criteria I have actually put forward, consider criterion (2): 'The relevant scientific community must incorporate a substantial diversity of perspectives.' Now clearly one couldn't ask for anything like an optimally diverse scientific community, since then the criterion would never be satisfied, not even for those 30 examples in Chapter 1; future-proof science would be impossible. Instead the criterion asks for a 'substantial' diversity of perspectives, which is just another way of asking for *sufficient* diversity. But isn't this inadmissibly vague? How are we ever to 'tick off' the criterion in our search for future-proof scientific claims?

My answer is that modern-day scientific communities actually *do* incorporate a diversity of perspectives sufficient for future-proof science, meaning that one needn't worry about this criterion when one considers a contemporary scientific claim. The diversity of perspectives is of course *far* from optimal, for example because there is nothing like gender 'balance' in many scientific communities. But on the other hand all contemporary scientific communities are international to a good degree, and have made headway on the issue of gender balance in recent decades, such that any bias in perspective due to the large percentage of men making up the relevant community is assuaged. To put it another way, there are many female scientists in each scientific community, even if the *percentage* of female scientists is sometimes embarrassingly low (see Huang, J et al. 2020 for some helpful statistics).

So there is some measure of diversity, but how do we know there is *enough*, specifically for us to be sure that pertinent scientific claims are future-proof? Why am I confident that the criterion is met for modern-day scientific communities? The answer comes from considering the communities that were operative vis-à-vis those 30 examples listed in Chapter 1 *when the 95 per cent consensus was first secured.* In all of those 30 examples the 95 per cent criterion was met decades ago. As I've argued, since each consensus formed there has been enormous opportunity for the view to be overturned, if it was ever going to be. Now if the diversity of the relevant community was *in*sufficient, all those decades ago, then it is highly likely that at least one of the claims would have been overturned by now. But that hasn't happened. And if the communities were diverse enough for future-proof science back *then*, then they certainly are *now*, given that scientific diversity has increased.

Thus a benchmark for 'sufficient diversity' would come from an investigation into the communities that were behind such clear examples of future-proof science. Ideally I would take the time to analyse the diversity of those communities, and on the basis of that analysis I could be much more explicit concerning what counts as sufficient diversity. But in practice this isn't necessary. Since the communities that formed those consensuses several decades ago did—it turns out—comprise a sufficient diversity of perspectives, we can have confidence that modern-day communities are sufficiently diverse. Scientific community diversity is far from ideal for various reasons, but the only issue here is whether it is sufficient for future-proof science in cases where criterion (1) is met.[6]

A final worry here is that the field of fundamental physics isn't yet diverse enough for future-proof science. The thought that other fields *are* sufficiently diverse—given that they were sufficiently diverse in the past (e.g. 1960, or whenever a 95 per cent consensus was originally reached), and are even more diverse now—doesn't show that the fundamental physics community is sufficiently diverse. The fundamental physics community is more diverse than it used to be, and it is certainly highly international, but its measure of (gender) diversity remains below that found in other scientific fields. However, the diversity within the fundamental physics community *today* is comparable with the diversity of other scientific fields *several decades ago*, and that degree of diversity *was* sufficient for future-proof science, as the 30 examples given in Chapter 1 testify. (The interested reader is directed to Huang, J et al. 2020 for helpful data and discussion on the specific question of gender balance. One can also find helpful discussion in Miller 2013, p. 1312f.)

2.3 Counterexamples

Some readers may feel that they can think of an example from the history of science that constitutes a counterexample, or a close counterexample, to my primary thesis that a claim getting through the filter will be future-proof. And a *close* counterexample would be worrying enough, since it would suggest that a true counterexample is possible. For example, if we can find a case where 90 per cent of the relevant scientific community were completely convinced of something, for a decent stretch of time, and then it was later overturned, then it would be reasonable to suppose that whatever factors brought about that

[6] The reasoning may appear circular here, but it isn't. I submit that the vast majority of readers will agree that my 30 examples are good examples of future-proof science. (I haven't attempted to prove that they meet my criteria for future-proof science, of course.) With this agreed, the examples can be used to gauge what counts as sufficient community diversity for future-proof science. Thanks to an anonymous reviewer for urging me to clarify this.

90 per cent consensus (misleading evidence? A bandwagon effect?) could, in another case, act to bring about a 95 per cent consensus.

Of course, the history of science is a very big place. I have been working in integrated history and philosophy of science for more than 15 years, and have thought seriously about all of the prominent historical cases I have come across in the literature. The journey started with the famous list of examples found in Laudan (1981), but I attempted to both extend and improve Laudan's list in Vickers (2013a). Since 2013, I have continued to come across new, challenging cases, and I have no doubt further challenging cases are out there, either in literature I have missed, or else still waiting to be discovered.

I can already briefly mention a couple of interesting cases here. The first concerns the claim that a 'hidden variables' version of quantum mechanics is impossible. Clearly this has been overturned, since we now have Bohmian mechanics as a 'hidden variables' interpretation of quantum mechanics that is taken very seriously by the relevant community. However, there was a time when the majority of the relevant community were sure that such an interpretation was impossible. The renowned 'genius' John von Neumann had presented a 'proof' that such an interpretation was impossible in 1932, in his book *The Mathematical Foundations of Quantum Mechanics* (von Neumann 1932), and this 'proof' was quickly and widely accepted. Feyerabend (1996) recalls a debate concerning one of Niels Bohr's public lectures (*c.*1950), a dispute between Bohr's followers and his critics. He writes as follows:

> The Bohrians did not clarify the arguments; they mentioned an alleged proof by von Neumann, and that settled the matter. Now I very much doubt that those who mentioned the proof, with the possible exception of one or two of them, could have explained it. I am also sure that their opponents had no idea of its details. Yet, like magic, the mere name 'von Neumann' and the mere word 'proof' silenced the objectors. (p. 77f.)

This illustrates the influence of von Neumann's attempted proof. More dramatically still, the flaw in von Neumann's 'proof' was exposed very quickly by Grete Hermann, but she was unable to halt the snowball von Neumann had set in motion, and the 'proof' was widely accepted for more than 30 years (Seevinck 2017). As Mermin (1993) writes, 'A few years later Grete Hermann... pointed out a glaring deficiency in the argument, but she seems to have been entirely ignored. Everybody continued to cite the von Neumann proof.'

Whilst this is a striking example of the 'misplaced trust in a figure of huge authority' problem, it isn't at all obvious that it meets *either one* of my criteria for future-proof science. For one thing, we're not just talking about physics here, but about foundations of physics. This was a very small, very male community in the 1930s, 1940s, and 1950s. Good data on the gender gap in this field in this period is hard to come by, but it is clear that female experts in the field comprised less than 10 per cent of the community, and probably less than 5 per cent.[7] It seems reasonable to suppose that had the community been more gender balanced, Grete Hermann would have had a better chance of getting her voice heard.

In addition, it is reasonable to doubt that the community consensus reached 95 per cent, or even something close. Within scientific communities, a big name such as 'John von Neumann' will influence many, but a significant percentage will completely dismiss the *name*—whoever it is—and instead demand to be convinced by the arguments and the (first-order!) evidence. Even Einstein made significant mistakes, after all, and all professional physicists are aware that Einstein's reasoning went astray sometimes. In fact, if this *were* going to happen—if a solid 95 per cent consensus regarding some claim were going to form for purely sociological reasons such as the bandwagon effect—I submit that it would have happened by now. The history of science affords ample opportunity for it to happen. Thus we don't really need to present *reasons* for why it would never happen (recall Chapter 5, Section 3). It is enough to note that it has *not* happened, whatever the reasons are (recall Chapter 5, Section 4).

I fully agree with Herzenberg (2008)—drawing on Gribbin (1995)—that scientists *can* accept an idea because 'everybody knows' that it is true, and because it is written in a famous book, without checking the facts for themselves. But note the words 'scientists can'. I submit that only *some* scientists are *sometimes* influenced in this way. This is a long way from agreeing that a large, international, diverse community could form a 95 per cent consensus regarding the claim in question.

In all this it's important to remember what I intend by the word 'consensus'. I am certainly *not* following Miller (2013) who is willing to say that there is a scientific consensus when scientists merely 'accept', and do not believe, the claim in question. By contrast, I am concerned with cases where individual

[7] See e.g. Huang, J et al. (2020, p. 3 of 8), who note that 'in 1955 women represented only 12% of all active [scientific] authors', but this lumps together all different scientific disciplines, including some—psychology, political science, and health science—where the percentage of female scientists is significantly higher than many other fields, such as mathematics and physics.

scientists within the community would be willing to state that we are dealing with an 'established scientific fact'. This raises the bar quite considerably compared with the way the word 'consensus' is used, and defined, in *some* of the literature.

The von Neumann case is hardly a one-off, but it may be the most striking example of this phenomenon. Back in Chapter 5, I briefly mentioned the case of Eddy Palmer's 'proof', in 1954, that peptic ulcer disease could *not* be caused by bacteria. Many years later it turned out that Palmer was wrong (Zollman 2010; Šešelja and Straßer 2014). This time the case can't so easily be dismissed on the grounds of a lack of community diversity. But Eddy Palmer's name was nowhere near as powerful as John von Neumann's name, and I submit that the 'consensus'—such as it was—never reached anything like 95 per cent.[8]

2.4 Is the Sun a Star?

In this book I have often appealed to the statement 'The Sun is a star' as an example of a scientific fact that is clearly beyond dispute, and that even most sceptics would want to accept. I am 100 per cent sure(!) that this statement meets my criteria for future-proof science, even without conducting any kind of survey of relevant scientists. However, it may be objected that I have moved too quickly: perhaps there *are* reasonably good reasons for holding back from declaring this statement 'future-proof'.[9] A really determined sceptic might insist that it would be dangerous to completely close off debate on this issue, and, if this is one of my best examples, perhaps that entails that it would be dangerous to close off debate on *any* of my examples. My 30 examples *might* be future-proof, says this sceptic, but we should not *declare* them to be future-proof. We shouldn't be so certain, or at least it is reasonable to be *un*certain.

The worry starts with the thought that 'The Sun is a star' might go the same way as 'Pluto is a planet'. Of course, there was a time when it was 'obvious' that Pluto was one of the planets (1970, let's say). Now we teach our children that Pluto is *not* a planet. The reason is due to a new way of thinking about the 'planet' category, drawn up by the International Astronomical Union (IAU) in

[8] In the final stages of completing this book, one reader suggested as a possible counterexample the late 20th-century 'neoclassical synthesis' in macroeconomics. This is certainly an interesting case to properly explore, but only a detailed case study could present a problem for the account put forward in this book. To start with, a specific proposition would need to be identified, as a candidate for the 95% consensus criterion.

[9] Many thanks to an anonymous reviewer for suggesting I think harder about this issue.

2006. When we embrace the 2006 definition, Pluto doesn't meet the criteria for a planet, because it doesn't 'clear its neighbourhood' (see e.g. Bokulich 2014). Why not apply this same thinking to 'The Sun is a star'? Clearly the scientific community is not averse to changing definitions of key terms from time to time, as new scientific understanding calls for it. Isn't it possible that, 500 years from now, the word 'star' will be defined in some new, special way that we can't even conceive of right now, and such that our own Sun does *not* meet the definition of 'star'?

An initial response notes that the two cases are importantly different. Pluto was always an outlier compared with the other planets, for example because its orbit is much more eccentric, and is significantly angled to the ecliptic. By contrast, '[O]ur Sun is very much a run-of-the-mill star' (Noyes 1982, p. 7). The fact that our Sun is such a *typical* star dramatically reduces the plausibility that it will one day go the same way as Pluto.

The sceptic might push back, however, insisting that perhaps 'star' won't even be considered a serious category 500 years from now. If we consider the concept of *jade*, this fragmented into two separate kinds, *jadeite* and *nephrite*, in the 19th century. The concept *jade* has no scientific significance, with the word 'jade' used merely as an umbrella term for jadeite and nephrite (cf. Bird and Tobin 2022). Similarly, whilst the word 'star' seems to refer to a single, scientifically serious kind *right now*, it might, in the future, be no more than an umbrella term for two or more 'fragmented' concepts (Taylor and Vickers 2017). So the argument goes.

But again, the two cases are significantly different. The kind known as 'jade' prior to the 1860s was never very well understood from a scientific point of view. By contrast, the kind known as 'star' is exceedingly well understood. We do of course recognise different kinds of star, including neutron stars, brown dwarf stars, and red supergiants such as Betelgeuse. But the overarching category 'star' is considered very scientifically significant, and its definition has been taken seriously in the literature (in a way the definition of 'planet' never was, prior to 2006). For example, in Prialnik (2009), *An Introduction to the Theory of Stellar Structure and Evolution*, Section 1.1 'What is a star?', one finds that, 'A star can be defined as a body that satisfies two conditions: (a) it is bound by self-gravity; (b) it radiates energy supplied by an internal source.' Our own Sun clearly meets this definition, along with (the objects that stand behind) the vast majority of those twinkly points of light in the night sky.

There's an important lesson here for all 30 of my examples of future-proof science listed in Chapter 1. If we are pressed to articulate what, precisely, each proffered future-proof statement means, a balance must be struck between saying too much and saying too little. On a very *thin* reading of 'star' it becomes trivial that the Sun is a star, because if we refuse to say anything substantial concerning what a star *is*, then, whatever happens in the future, we will be able to insist that the Sun is a star. But on a very *thick* reading of 'star', where we say *a lot* about what a star is, it becomes less obvious that 'The Sun is a star' is future-proof: the more we say the more likely it is that something, somewhere, will change in the future. And if the bare *sentence* 'The Sun is a star' is retained, but its meaning changes significantly, then (as previously discussed) that isn't the kind of 'future-proof science' this book aims to identify. It is purely verbal continuity, whilst the underlying scientific ideas vary.

But as the definition of 'star' found in Prialnik (2009) demonstrates, it isn't too hard to find a happy medium. Prialnik's definition is highly non-trivial (and so doesn't say too little), but also doesn't include any remotely 'risky' content from the point of view of epistemology (it doesn't say too much). I submit that there is no doubt in the relevant scientific community that our own Sun meets Prialnik's basic textbook definition of 'star', and so too do the *vast* majority of the objects that give rise to those twinkly lights in the night sky. My reference to a 'basic textbook definition' is important here as I generalise this point across all 30 of the examples of future-proof science listed back in Chapter 1. If one takes any one of these examples, and asks 'How *thin* or *thick* is this claim?' or simply 'What does this really mean?', my answer will draw on basic textbook definitions of the key terms.

This is a rather brief response—there are of course many rich debates concerning concepts, meaning, reference, and 'natural kinds'—but I hope it is suitably clarificatory for present purposes.

3. Implications for School Education

If the message at the heart of this book is right, then a shift in science education is called for. The argument starts with the fact that many of us want children to leave school with the best possible tools for navigating various life choices. Supposing we care about climate change, air pollution, vaccination against a range of diseases, condom use in the battle against HIV, society behaviour during a pandemic, and so on, then we will also care that people make appropriate life choices relevant to these values. And of course, any doubts about climate change, the safety of the MMR vaccine, and so on, will likely lead to inappropriate choices. Some would then say that the best way to tackle this is to ensure that children leave school knowing as much science as possible. They should be taught about biology, chemistry, and physics, they should be taught how to do scientific experiments, how to use microscopes effectively, and so on. If Enfield (2008, p. 891) is right that '[t]he only sensible way to find out whether a scientific claim is true is by looking at the original scientific evidence previously offered by scientists', then we need people quite generally to acquire the ability, primarily through school education, to digest and assess that original scientific evidence.

But, according to this book, there is a serious flaw in this line of thinking. As argued, the best way to judge whether or not to believe a given scientific claim is really *not* to attempt to digest and assess the original scientific evidence for the claim. Even the most able teenagers, leaving school at 16 or 18 with top marks in their science exams, could *not* hope to be able to competently assess the original scientific evidence for the vast majority of scientific claims they may come to care about. To give just one concrete example, a good student might be able to read and understand Carol Cleland's excellent summary of the reasons for believing that a bolide (asteroid) impact caused the dinosaur extinction (Cleland 2013, pp. 4–5). This would stand in place of reading hundreds of original journal articles, which of course is too much to ask of any layperson. Based on Cleland's summary, the student could easily be convinced; it *seems* so convincing. But that summary, however well done, however apparently convincing, shouldn't persuade anyone. That same year a paper came out in *Geological Magazine* arguing that the Chicxulub bolide impact occurred more than 100,000 years prior to the mass extinction (Keller et al. 2013). The authors of that paper were professional scientists, much more knowledgeable regarding the evidence in favour of the Alvarez impact hypothesis than any school-leaver ever could be. And yet they explicitly deny the story told in the Cleland (2013) paper. What seems so convincing is actually full of holes.

More generally, no school-leaver could realistically, competently assess the first-order scientific evidence for any claim they care about. Any attempt to do so could only result in an extremely impoverished sense of the overall evidence, and might well be thoroughly biased, missing out lots of important considerations. Nor could the school-leaver read a good *summary* or *review* of the overall evidence and think that that was sufficient. Any such summary or review would have to skip over a huge number of important details, and might itself be biased in one way or another. After all, most such summaries are written by scientists (who do know some reasonable fraction of the first-order evidence), and scientists tend to have their biases, based on their academic training, nationality, precise specialism, and so on. Those with similar biases also tend to group together, so the fact that a review article (e.g. Schulte et al. 2010, as discussed in Chapter 7) has 40 co-authors is no indication that it is not biased.

This book argues that there is another way. Instead of putting all of the emphasis on teaching schoolchildren about *science*, we could put a far greater emphasis on teaching schoolchildren about *scientific communities*, and in particular the features of those communities that correlate strongly with trustworthy scientific claims.[10] Specifically, schoolchildren could be taught how to competently judge when a solid scientific consensus does/doesn't exist vis-à-vis some claim of interest. Older teenagers would then leave school less able to work a microscope, perhaps, but *more* able to digest online information relating to scientific community opinion. For example, they would be able to clearly distinguish between the case of the Alvarez hypothesis, and the case of anthropogenic climate change, *not* by seeing directly that the evidence for anthropogenic climate change is stronger, but by identifying a significant (though small) scientific community opposition to the former, but not the latter. More generally, they would be better equipped to distinguish between cases of mere *majority* opinion (e.g. the fixists about continents in the 1930s, 1940s, and 1950s), and cases of solid scientific consensus. They would leave school with a richer concept of 'scientific consensus', and also a richer concept of 'scientific community', including the senses in which it is reasonable to ascribe epistemic opinions to a community (as opposed to an individual) in the first place. Whilst the details of how exactly to modify school curricula would need to be worked out, the direction of travel recommended by this book is clear.

I am certainly not the first to make such claims about scientific education. Consider Allchin (2020) discussing the Covid-19 pandemic:

> A major goal in education has typically been intellectual independence for all. Namely, if everyone can evaluate arguments and evidence themselves, then shouldn't science triumph? But, ironically, such an attitude fosters an illusion of competence that displaces deference to expertise. [...] Science students may need to learn, instead, how to cope with intellectual humility and when to respect expert consensus. [...] The verdict of recent COVID-19

[10] Cf. Dunlop and Veneu (2019, especially Section 4.2), who also argue that school education systems need to step away from merely 'teaching the facts', making room for lessons about scientific community dynamics, although they specifically focus on the possibility of teaching children about scientific *controversies*.

> history seems to be this: Rely on the experts, not one's own fragile personal assessment of the evidence. (p. 432)

Precisely how one should 'rely on the experts' is no easy question. There has been important literature on key concepts in this story, including *consensus* (e.g. Miller 2013) and *expertise* (e.g. Collins and Evans 2008). The relationship between expertise, consensus, and public decision-making is a hot topic (e.g. Goldenberg 2021). In this book I've tried to add something to such literature. Specifically, my hope is that it is clearer than ever before that a significant intervention in science education programmes around the world is called for, on the grounds that our children really do need to leave school with richer conceptions of expertise, consensus, and scientific community dynamics (even if that means teaching them less physics, say). In this way I believe we can significantly improve the relationship between scientific communities and the wider public, at least for the *next* generation, who are just now entering school education systems around the world.

4. Outlook

This book does a lot in some ways, and, of course, not so much in others. In closing, I will mention some obvious next steps that I personally find attractive—though the reader may prefer others.

One obvious next step was highlighted in the previous section. Supposing this book is right to maintain that a shift in emphasis in science education in our schools—away from science, and towards scientific *communities*—would better equip school-leavers with the skills they need to navigate life decisions relating to science. There remains quite a gulf to bridge between this academic monograph and real-world policy change. Steps towards constructing this bridge would be an obvious extension of this project. An intriguing sub-project would compare alternative approaches found within the school education systems of different countries (see the journal *Comparative Education*). Which countries (if any) *already* put significant emphasis on the study of scientific communities? And can we measure the impact of such alternative approaches?

Another extension, more focused on the academic arguments, concerns venturing much deeper into social epistemology. No doubt some readers will wish there had been greater emphasis on social epistemology and sociology of science generally, with corresponding engagement with the relevant literature, including articles published in journals such as *Social Epistemology* and

Episteme. But of course, one could always do more in a book. Such a venture into the guts of social epistemology is left for the sequel. Crucial questions that might be explored include the following:

- How should we measure scientific community diversity? Can diversity along one dimension (e.g. nationality) make up for a lack of diversity along another dimension (e.g. gender)?
- How do we navigate the language of scientists when trying to judge whether or not they *would* be willing (if asked) to describe a claim as an 'established scientific fact'?
- How should we compare situations where a large majority of the community are *quite* sure, with situations where a smaller majority are *certain*?
- When we design large-scale surveys of scientific opinion, how do we best ensure that they are reliable guides to the overall opinion of the entire, relevant scientific community?

As we tackle such questions, important literature to engage with includes Longino (1990, 2002), Tucker (2003), Tollefsen (2006), Wylie (2006), Fuller (2007), Feldman (2009), Shwed and Bearman (2010), Anderson (2011), Miller (2013), Solomon (2001, 2015), Stegenga (2016), and Beatty (2017). For now (with apologies) I merely direct the interested reader to this literature. I would add, though, that much of this literature strikes me as only partially relevant, given that these authors are *not* focused on *identifying future-proof science*. Consider the way Miller (2013) characterises 'consensus', for example. He writes that 'consensus is a species of acceptance, rather than proper belief. Acceptance only requires that one take certain claims for granted for the sake of reasoning, without necessarily taking them to be true' (p. 1303). But in this book, when I use the term 'consensus', I am referring to cases where individual scientists do not *merely* 'accept' a claim 'for the sake of reasoning'. Thus Miller's work, although appearing directly relevant to my project, is actually quite tangential in important respects.

When we specifically ask about future-proof science, old questions take on new significance, the dialectical landscape is adjusted, and there is much to explore. Moving forward, I am hopeful that others will help me explore that landscape. But whilst I think there is value in exploring a landscape, something bothers me here. Clearly what's driven this entire project is a desire to identify *facts*. And the proposal is that we should identify facts via a very strong consensus (second-order, not first-order, evidence). Now we may ask: Why shouldn't these claims about identifying facts apply just as much to claims in *philosophy*? Assume that's right—assume there isn't some *special* way of reaching 'the facts' in philosophy that separates it from science. So, we should identify factual claims in philosophy via second-order evidence. Now, what I am putting forward in this book is essentially a philosophy of science. And, since strong consensuses in philosophy are vanishingly rare (for positive claims, at least[11]), I shouldn't even remotely hope that one day it will be reasonable to describe these ideas about identifying future-proof science as *true* or *factual*. Strictly speaking, I shouldn't believe what I am saying in this book, at least not until a consensus of relevant experts (science studies scholars, broadly construed) emerges. And it seems likely that that will never happen.

Perhaps with some degree of irrationality, I remain optimistic about consensuses in philosophy. There is a clear culture of critical thinking in philosophy, and it is not unusual for journal articles to emerge that do nothing more than criticise an idea somebody else has put forward, finding the gap in the argument, or the weak premise. And of course, one can't hope to publish a paper that merely *agrees* with an idea that has already been published. Thus when one looks to the literature one will see all sorts of *dis*agreement, and perhaps little agreement, but this needn't mean that there *aren't* any interesting points of community agreement. It's just less obvious how exactly that agreement makes it into the literature. Perhaps making points of widespread agreement explicit is something our community needs to work harder on, and shouldn't expect to happen naturally, just as the scientific community sometimes needs to *make a special effort* to determine how strongly they believe in something.

In the meantime, I will remain conflicted, believing the core ideas put forward in this book on a Friday (observing from the inside, wearing my philosophy hat), and withholding belief on a Saturday (looking on from the outside, wearing an unfamiliar sociology hat). Previously I commented that we must hide this book from the scientists, since we want them to continue assessing scientific ideas *from the inside*, using first-order evidence, even if second-order evidence is a better guide to truth. Now it seems that I should hide it *from myself*. A good first step in that direction is to stop writing it.

[11] There may well be a consensus that knowledge is *not* justified true belief, for example.

Bibliography

Abrams M (2007), 'How do natural selection and random drift interact?'. *Philosophy of Science* **74**: 666–79.

Adelmann HB (1966), *Marcello Malpighi and the Evolution of Embryology* (Ithaca, NY: Cornell University Press).

Adie RJ (1952), 'The position of the Falkland Islands in a reconstruction of Gondwanaland'. *Geol. Mag.* **89**: 401–10.

Agharkar SP (1937), 'Wegener's theory of continental drift with reference to India and adjacent countries (general discussion)'. *Proceedings 24th Indian Science Congress*, Hyderabad: 502–20.

Ahlberg E and Clack JA (2006), 'A firm step from water to land'. *Nature* **440**: 747–9.

Alai M (2017), 'Resisting the historical objections to realism: is Doppelt's a viable solution?'. *Synthese* **194**(9): 3267–90.

Allchin D (2020), 'The Covid-19 conundrum'. *The American Biology Teacher* **82**(6): 429–33.

Allzén S (2022), 'From Unobservable to Observable: Scientific Realism and the Discovery of Radium', *Journal for General Philosophy of Science*, forthcoming.

Almécija S, Hammond AS, Thompson NE, Pugh KD, Moyà-Solà S, and Alba DM (2021), 'Fossil apes and human evolution'. *Science* **372**(6542): eabb4363.

Almeida JD, Berry DM, Cunningham CH, Hamre D, Hofstad MS, Mallucci L, McIntosh K, and Tyrrell DAJ (1968), 'Virology: coronaviruses'. *Nature* **220**(5168): 650.

Alvarez LW, Alvarez W, Asaro F, and Michel HV (1980), 'Extraterrestrial cause for the Cretaceous-Tertiary Extinction'. *Science* **208**(4448): 1095–108.

Anderson E (2011), 'Democracy, public policy, and lay assessments of scientific testimony'. *Episteme* **8**(2): 144–64.

Archibald JD, Clemens WA, Padian K, Rowe T, MacLeod N, Barrett PM, Gale A, Holroyd P, Sues H, Arens NC, Horner JR, Wilson GP, Goodwin MB, Brochu CA, Lofgren DL, Hurlbert SH, Hartman JH, Eberth DA, Wignall PB, Currie PJ, Weil A, Prasad GVR, Dingus L, Courtillot V, Milner A (Angela), Milner A (Andrew), Bajpai S, Ward DJ, and Sahni A (2010), 'Cretaceous extinctions: multiple causes'. *Science* **328**: 973.

Argand E (1924), 'La tectonique de l'Asie'. *Proceedings of the XIIIth International Geological Congress* **1** (part 5): 171–372.

Arndt M, and Hornberger K (2014), 'Testing the limits of quantum mechanical superpositions'. *Nature Physics* **10**: 271–7.

Arndt M, Nairz O, Vos-Andreae J, Keller C, van der Zouw G, and Zeilinger A (1999), 'Wave-particle duality of C_{60} molecules'. *Nature* **401**: 680–2.

Asay J (2019), 'Going local: a defense of methodological localism about scientific realism'. *Synthese* **196**: 587–609.

Bailey EB (1928), 'The Palaeozoic mountain systems of Europe and America'. *Nature* **122**: 811–14.

Barnes EC (2008), *The Paradox of Predictivism* (Cambridge: Cambridge University Press).

Beatty J (2017), 'Consensus: sometimes it doesn't add up'. Chapter 15 in SB Gissis, E Lamm, and A Shavit (eds), *Landscapes of Collectivity in the Life Sciences* (Cambridge, MA: MIT Press): 179–98.

Bergin AE and Strupp HH (1972), *Changing Frontiers in the Science of Psychotherapy* (Chicago, IL: Aldine-Atherton).

Berry DM, Cruickshank JG, Chu HP, and Wells RJ (1964), 'The structure of infectious bronchitis virus'. *Virology* **23**(3): 403–7.

Best NW (2015), 'Lavoisier's "Reflections on phlogiston" I: against phlogiston theory', *Foundations of Chemistry* **17**(2): 137–51.

Biedenharn LC (1983), 'The "Sommerfeld Puzzle" revisited and resolved'. *Foundations of Physics* **13**(1): 13–34.

Bigarella JJ and Salamuni R (1961), 'Early Mesozoic wind patterns as suggested by dune bedding in botucatú sandstone of Brazil and Uruguay'. *Bull. Geol. Soc. Am.* **72**: 1089.

Bird A (2007), *Nature's Metaphysics: Laws and Properties* (Oxford: Clarendon Press).

Bird A (2018), 'Understanding the replication crisis as a base-rate fallacy'. *British Journal for the Philosophy of Science* **72**(4): axy051.

Bird A (2022): *Knowing Science* (Oxford: Clarendon Press).

Bird A and Tobin E (2022), 'Natural kinds'. *The Stanford Encyclopedia of Philosophy* (Spring 2022 Edition), ed. Edward N. Zalta; available at https://plato.stanford.edu/archives/spr2022/entries/natural-kinds/.

Böhme M, Spassov N, Fuss J, Tröscher A, Deane AS, Prieto J, Kirscher U, Lechner T, and Begun DR (2019), 'A new Miocene ape and locomotion in the ancestor of great apes and humans'. *Nature* **575**: 489–93.

Bokulich A (2014), 'Pluto and the "Planet Problem": folk concepts and natural kinds in astronomy'. *Perspectives on Science* **22**(4): 464–90.

Bolinska A, and Martin JD (2021), 'The tragedy of the canon; or, path dependence in the history and philosophy of science'. *Studies in History and Philosophy of Science* **89**: 63–73.

Boudry M (2020), 'A strange paradox: the better we manage to contain the coronavirus pandemic, the less we will learn from it'. *The Conversation* (2 April 2020); available at https://theconversation.com/a-strange-paradox-the-better-we-manage-to-contain-the-coronavirus-pandemic-the-less-we-will-learn-from-it-135268.

Brusatte, SL (2015), 'What killed the dinosaurs?'. *Scientific American* (December 2015): 54–9.

Brusatte SL, Butler RJ, Barrett PM, Carrano MT, Evans DC, Lloyd GT, Mannion PD, Norell MA, Peppe DJ, Upchurch P, and Williamson TE (2015), 'The extinction of the dinosaurs'. *Biological Reviews* **90**: 628–42.

Brush SG (2015), *Making 20th Century Science: How Theories Became Knowledge* (New York: Oxford University Press).

Bryant A (2020), 'Naturalisms'. *Think* **19**(56): 35–50.

Bullard E (1975), 'The emergence of plate tectonics: a personal view'. *Annual Review of Earth and Planetary Sciences* **3** (May 1975): 1–31.

Burdett IDJ, and Rogers HJ (1970), 'Modification of the appearance of mesosomes in sections of *Bacillus licheniformis* according to the fixation procedures'. *Journal of Ultrastructure Research* **30**: 354–67.

Burian RM (1997), 'Exploratory experimentation and the role of histochemical techniques in the work of Jean Brachet, 1938–1952'. *History and Philosophy of the Life Sciences* **19**: 27–45.

Callender C (2020), 'Can we quarantine the quantum blight?'. Chapter 4 in S French and J Saatsi (eds), *Scientific Realism and the Quantum* (Oxford: Oxford University Press).

Camilleri K (2007), 'Indeterminacy and the limits of classical concepts: the transformation of Heisenberg's thought'. *Perspectives on Science* **15**(2): 178–201.

Campbell, DH (1944), 'Living fossils'. *Science* **100**: 179–81.

Cantor G (1975), 'The reception of the wave theory of light in Britain: a case study illustrating the role of methodology in scientific debate'. *Historical Studies in the Physical Sciences* **6**: 109–32.

Carey SW (1955), 'The orocline concept in geotectonics'. *Roy. Soc. Tasmania* **89**: 255–88.

Carrier M (2004), 'Experimental success and the revelation of reality: the miracle argument for scientific realism'. In M Carrier et al. (eds), *Knowledge and the World: Challenges Beyond the Science Wars* (Heidelberg: Springer): 137–61.

Cartwright N, Hardie J, Montuschi E, Soleiman M, and Thresher A (forthcoming), *The Tangle of Science: Reliability Beyond the Scientific Method, Rigour, and Objectivity* (Oxford: Oxford University Press).

Chakravartty A (2021), 'Realist representations of particles: the Standard Model, top down and bottom up'. Chapter 15 in TD Lyons and P Vickers (eds), *Contemporary Scientific Realism: The Challenge from the History of Science* (New York: Oxford University Press).

Chalmers A (2009), *The Scientist's Atom and the Philosopher's Stone: How Science Succeeded and Philosophy Failed to Gain Knowledge of Atoms* (Boston Studies in the Philosophy and History of Science) (Dordrecht: Springer).

Chang H (2004), *Inventing Temperature: Measurement and Scientific Progress* (New York: Oxford University Press).

Chiarenza AA, Farnsworth A, Mannion PD, Lunt DJ, Valdes PJ, Morgan JV, and Allison PA (2020), 'Asteroid impact, not volcanism, caused the end-Cretaceous dinosaur extinction'. *PNAS* **117**(29): 17084–93.

Churchill FB (1991), 'The rise of classical descriptive embryology'. In SF Gilbert (ed.), *Development Biology: A Comprehensive Synthesis: Volume 7: A Conceptual History of Modern Embryology* (New York: Plenum Press): 1–29.

Churchland PM and Hooker CA (eds) (1985), *Images of Science: Essays on Realism and Empiricism, with a Reply from B.C. van Fraassen* (Chicago, IL: University of Chicago Press).

Cleland CE (2013), 'Common cause explanation and the search for a smoking gun'. In VR Baker (ed.), *Rethinking the Fabric of Geology* (Geological Society of America Special Paper 502) (Boulder, CO: Geological Society of America): 1–9.

Cleland CE (2019), *The Quest for a Universal Theory of Life: Searching for Life as We Don't Know It* (Cambridge: Cambridge University Press).

Collins H and Evans R (2008), *Rethinking Expertise* (Chicago, IL: University of Chicago Press).

Cook J, Nuccitelli D, Green SA, Richardson M, Winkler B, Painting R, Way R, Jacobs P, and Skuce A (2013), 'Quantifying the consensus on anthropogenic global warming in the scientific literature'. *Environmental Research Letters* **8**, 024024: 1–7.

Cook J, Oreskes N, Doran PT, Anderegg WRL, Verheggen B, Maibach EW, Carlton JS, Lewandowsky S, Skuce AG, Green SA, Nuccitelli D, Jacobs P, Richardson M, Winkler B, Painting R, and Rice K (2016), 'Consensus on consensus: a synthesis of consensus estimates on human-caused global warming'. *Environmental Research Letters* **11**, 048002: 1–7.

Courtillot V (1999), *Evolutionary Catastrophes: The Science of Mass Extinction* (Cambridge: Cambridge University Press) [English translation of the original French edition (1995)].

Courtillot V and Fluteau F (2010), 'Cretaceous extinctions: the volcanic hypothesis'. *Science* **328**: 973–4.

Culp S (1994), 'Defending robustness: the bacterial mesosome as a test case'. In D Hull, M Forbes, and RM Burian (eds), *Proceedings of the Biennial Meeting of the Philosophy of Science Association 1994*, Vol. 1 (Dordrecht: Reidel): 46–57.

Curnow T (2012), *A Practical Guide to Philosophy for Everyday Life: See the Bigger Picture* (London: Icon Books).

Cushing JT (1991), 'Quantum theory and explanatory discourse: endgame for understanding?'. *Philosophy of Science* **58**(3): 337–58.

Daly RA (1926), *Our Mobile Earth* (New York: Charles Scribner's Sons).

Dardashti R (2019), 'Physics without experiments?'. Chapter 9 in R Dardashti, R Dawid, and K Thébault (eds), *Why Trust a Theory? Epistemology of Fundamental Physics* (Cambridge: Cambridge University Press): 154–72.

Dardashti R (2021), 'No-go theorems: what are they good for?'. *Studies in History and Philosophy of Science Part A* **86**(April 2021): 47–55.

Daumer M, Dürr D, Goldstein S, and Zanghi N (1997), 'Naïve realism about operators'. *Erkenntnis* **45**: 379–97.

David-Ferreira JF and Manaker RA (1965), 'An electron microscope study of the development of a mouse hepatitis virus in tissue culture cells'. *The Journal of Cell Biology* **24**(1): 57–78.

De Solla Price DJ (1963), *Little Science, Big Science* (New York: Columbia University Press).

Dellsén F (2017), 'Reconstructed empiricism'. *Acta Analytica* **32**: 95–113.

DeMille D (2015), 'Diatomic molecules, a window onto fundamental physics'. *Physics Today* **68**(12): 34–40.

DePalma RA, Smit J, Burnham DA, Kuiper K, Manning PL, Oleinik A, Lerson P, Maurrasse FJ, Vellekoop J, Richards MA, Gurche L, and Alvarez W (2019), 'A seismically induced onshore surge deposit at the KPg boundary, North Dakota'. *Proceedings of the National Academy of Sciences* **116**(17): 8190–9.

Díez JA and Carman C (2015), 'Did Ptolemy make novel predictions? Launching Ptolemaic astronomy into the scientific realism debate'. *Studies in the History and Philosophy of Science* **52**: 20–34.

Dixey F (1939), 'Some observations on the physiographical development of central and southern Africa'. *Trans. Proc. Geol. Soc. S. Afr.* **41**: 113–71.

Domermuth CH and Edwards OF (1957), 'An electron microscope study of chorioallantoic membrane infected with the virus of avian infectious bronchitis'. *The Journal of Infectious Diseases* **100**(1): 74–81.

Dorling J (1992), 'Bayesian conditionalization resolves positivist/realist disputes'. *Journal of Philosophy* **89**(7): 362–82.

Doumani GA and Long WE (1962), 'The ancient life of the Antarctic'. *Sci. Am.* **207**: 163–84.

Drosten C, Günther S, Preiser W, van der Werf S, Brodt H, Becker S, Rabenau H, Panning M, Kolesnikova L, Fouchier RAM, Berger A, Burguière A, Cinatl J, Eickmann M, Escriou N, Grywna K, Kramme S, Manuguerra J, Müller M, Rickerts V, Stürmer M, Vieth S, Klenk H, Osterhaus ADME, Schmitz H, and Doerr HW (2003), 'Identification of a novel coronavirus in patients with severe acute respiratory syndrome'. *New England Journal of Medicine* **348**: 1967–76.

Du Toit A (1928), 'Some reflections upon a geological comparison of South Africa with South America'. *Proceedings Geological Society South Africa*, Presidential Address (1928): 19–38.

Duhem P (1906), *La Théorie Physique, son objet et sa structure* (Paris: Chevalier et Rivière). English translation by P Wiener, *The Aim and Structure of Physical Theory* (Princeton, NJ: Princeton University Press, 1954).

Dunlop L and Veneu F (2019), 'Controversies in science: to teach or not to teach?'. *Science and Education* **28**: 689–710.

Egg M (2021), 'Quantum ontology without speculation'. *European Journal for Philosophy of Science* **11**: article 32, 1–26.

Eibenberger S, Gerlich S, Arndt M, Mayor M, and Tüxen J (2013), 'Matter–wave interference of particles selected from a molecular library with masses exceeding 10000 amu'. *Phys. Chem. Chem. Phys.* **15**: 14696–700.

Elga A (2010), 'How to disagree about how to disagree'. Chapter 7 in R Feldman and TA Warfield (eds), *Disagreement* (Oxford: Oxford University Press): 175–86.

Elliott KC (2007), 'Varieties of exploratory experimentation in nanotoxicology'. *History and Philosophy of the Life Sciences* **29**: 311–36.

Elsdon-Baker F (2009), *The Selfish Genius: How Richard Dawkins Rewrote Darwin's Legacy* (London: Icon Books).

Enfield P (2008), 'Book review of Stanford PK (2006), *Exceeding Our Grasp* (Oxford: Oxford University Press)'. *British Journal for the Philosophy of Science* **59**: 881–95.

Engstrom EJ (2016), 'Tempering madness: Emil Kraepelin's research on affective disorders'. *Osiris* **31**: 163–80.

Evenson KM, Wells JS, Petersen FR, Danielson BL, Day GW, Barger RL, and Hall JL (1972), 'Speed of light from direct frequency and wavelength measurements of the methane-stabilized laser'. *Phys. Rev. Lett.* **29**(19): 1346.

Fahrbach L (2011), 'How the growth of science ends theory change'. *Synthese* **180**: 139–55.

Fahrbach L (2017), 'Scientific revolutions and the explosion of scientific evidence'. *Synthese* **194**(12): 5039–72.

Fahrbach L (2021), 'We think, they thought: a critique of the pessimistic meta-meta-induction'. Chapter 13 in TD Lyons and P Vickers (eds), *Contemporary Scientific Realism: The Challenge from the History of Science* (New York: Oxford University Press).

Feldman R (2009), 'Evidentialism, higher-order evidence, and disagreement'. *Episteme* **6**(3): 294–312.

Fermor LL (1951), 'The mineral deposits of Gondwanaland'. *Trans. Inst. Min. Metall.* **60**: 421–65.

Feuer MJ (2006), *Moderating the Debate: Rationality and the Promise of American Education* (Cambridge, MA: Harvard Education Press).

Feyerabend P (1996), *Killing Time: The Autobiography of Paul Feyerabend* (Chicago, IL: University of Chicago Press).

Fine A (1986), 'Unnatural attitudes: realist and instrumentalist attachments to science'. *Mind* **95** (378): 149–79.

Fitz-James PC (1960), 'Participation of the cytoplasmic membrane in the growth and spore formation of bacilli'. *Journal of Biophysical and Biochemical Cytology* **8**: 507–28.

Fitzpatrick S (2013), 'Doing Away with the No Miracles Argument', in D Dieks & V Karakostas (eds.), *Recent Progress in Philosophy of Science: Perspectives and Foundational Problems* (Springer 2013), pp. 141–52.

Font E, Adatte T, Sial AN, Lacerda LD, Keller G, and Punekar J (2016), 'Mercury anomaly, Deccan volcanism, and the end-Cretaceous mass extinction'. *Geology* **44**(2): 171–4.

Font E, Carlut J, Rémazeilles C, Mather TA, Nédélec A, Mirão J, and Casale S (2017), 'End-Cretaceous akaganéite as a mineral marker of Deccan volcanism in the sedimentary record'. *Scientific Reports* **7**: 11453.

Frankel HR (2012), *The Continental Drift Controversy* (Cambridge: Cambridge University Press).

Franklin LR (2005), 'Exploratory experiments'. *Philosophy of Science* **72**: 888–99.

Fraser JD and Vickers P (2025), 'Knowledge of the Quantum Domain: An Overlap Strategy', *British Journal for the Philosophy of Science*, **76**.

Friedman M (2008), 'The evolutionary origin of flatfish asymmetry'. *Nature* **454**(7201): 209–12.

Froome, KD (1958), 'A new determination of the free-space velocity of electromagnetic waves'. *Proc. R. Soc. Lond. A*, **247**(1248): 109–22.

Frost-Arnold G (2019), 'How to be a historically motivated anti-realist: the problem of misleading evidence'. *Philosophy of Science* **86**(5): 906–17.

Fukuoka H, Sagawa T, Inoue Y, Takahashi H, and Ishijima A (2014), 'Direct imaging of intracellular signaling components that regulate bacterial chemotaxis'. *Science Signaling* **7**(319); http://doi.org/10.1126/scisignal.2004963.

Fuller S (2007), *The Knowledge Book: Key Concepts in Philosophy, Science and Culture* (Montreal: McGill-Queen's University Press).

Gardner MR (1979), 'Realism and instrumentalism in 19th-century atomism'. *Philosophy of Science* **46**(1): 1–34.

Gentile F, Moretti M, Limongi T, Falqui A, Bertoni G, Scarpellini A, Santoriello S, Maragliano L, Zaccaria RP, and di Fabrizio E (2012), 'Direct imaging of DNA fibers: the visage of double helix'. *Nano Letters* **12**(12): 6453–8.

Glymour C (2008), 'The paradox of predictivism'. *Notre Dame Philosophical Reviews* (13 June 2008) [review of Barnes EC (2008), *The Paradox of Predictivism* (Cambridge: Cambridge University Press)].

Goderis S, Sato H, Ferrière L, Schmitz B, Burney D, Kaskes P, Vellekoop J, Wittmann A, Schulz T, Chernonozhkin SM, Claeys P, de Graaff SJ, Déhais T, de Winter NJ, Elfman M, Feignon J, Ishikawa A, Koeberl C, Kristiansson P, Neal CR, Owens JD, Schmieder M, Sinnesael M, Vanhaecke F, Van Malderen SJM, Bralower TJ, Gulick SPS, Kring DA, Lowery CM, Morgan JV, Smit J, Whalen MT, IODP-ICDP Expedition 364 Scientists (2021), 'Globally distributed iridium layer preserved within the Chicxulub impact structure'. *Science Advances* **7**: eabe3647, 1–13.

Godfrey-Smith P (2003), *Theory and Reality* (Chicago, IL: University of Chicago Press).

Godfrey-Smith P (2008), 'Recurrent transient underdetermination and the glass half full'. *Philosophical Studies* **137**: 141–8.

Godfrey-Smith P (2021), *Theory and Reality*, 2nd edition (Chicago, IL: University of Chicago Press).

Goldenberg M (2021), *Vaccine Hesitancy: Public Trust, Expertise and the War on Science* (Pittsburgh, PA: University of Pittsburgh Press).

Good R (1947), *The Geography of the Flowering Plants* (London: Longmans, Green & Co.).

Gooday G and Mitchell D (2013), 'Rethinking "Classical Physics"'. Chapter 24 in JZ Buchwald and R Fox (eds), *The Oxford Handbook of the History of Physics* (Oxford: Oxford University Press): 721–64.

Gould SJ (1977), *Ontogeny and Phylogeny* (Cambridge, MA: Belknap Press of Harvard University Press).

Granovski YI (2004), 'Sommerfeld formula and Dirac's theory'. *Physics–Uspekhi*, **47**(5): 523–4.

Greenberg DS (1999), *The Politics of Pure Science*, New Edition (Chicago: University of Chicago Press).

Gribbin J (1995), *Schrödinger's Kittens and the Search for Reality* (Boston, MA: Little, Brown & Co.).

Griffiths DJ (2004), *Introduction to Quantum Mechanics*, 2nd edition (London: Pearson).

Grochala W (2018), 'On the position of helium and neon in the Periodic Table of Elements'. *Found Chem* **20**: 191–207.

Grocholski B (2019), 'A cooler core for Earth'. *Science* **364**(6439): 448–9.

Gulick SPS, Bralower TJ, Ormö J, Hall B, Grice K, Schaefer B, Lyons S, Freeman KH, Morgan JV, Artemieva N, Kaskes P, de Graaff SJ, Whalen MT, Collins GS, Tikoo SM,

Verhagen C, Christeson GL, Claeys P, Coolen MJL, Goderis S, Goto K, Grieve RAF, McCall N, Osinski GR, Rae ASP, Riller U, Smit J, Vajda V, and Wittmann A (2019), 'The first day of the Cenozoic'. *PNAS* **116**(39): 19342–51.

Hacking I (1983), *Representing and Intervening: Introductory Topics in the Philosophy of Natural Science* (Cambridge: Cambridge University Press).

Hafer A (2016), *The Not-So-Intelligent Designer: Why Evolution Explains the Human Body and Intelligent Design Does Not* (Cambridge: The Lutterworth Press).

Hallam A (1973), *A Revolution in the Earth Sciences* (Oxford: Clarendon Press).

Hallam A (1983), *Great Geological Controversies* (Oxford: Oxford University Press).

Hamilton WB (1961), 'Origin of the Gulf of California'. *Bull. Geol. Soc. Am.* **72**: 1307–18.

Hamilton WB (2002), 'The closed upper-mantle circulation of plate tectonics'. In S Stein and JT Freymueller (eds), *Plate Boundary Zones: Geodynamics Series*, Volume 30 (Washington, DC: American Geophysical Union): 359–410.

Hamre D, Kindig DA, and Mann J (1967), 'Growth and intracellular development of a new respiratory virus'. *Journal of Virology* **1**(4): 810–16.

Hardwig J (1991), 'The role of trust in knowledge'. *Journal of Philosophy* **88**: 693–708.

Hargather MJ, Settles GS, and Madalis MJ (2010), 'Schlieren imaging of loud sounds and weak shock waves in air near the limit of visibility'. *Shock Waves* **20**: 9–17.

Heilbron J and Kuhn T (1969), 'The genesis of the Bohr atom'. *Historical Studies in the Physical Sciences* **1**: 211–90.

Heisenberg W (1958), *Physics and Philosophy* (London: Allen & Unwin).

Heisenberg W (1968), 'Ausstrahlung von Sommerfelds Werk in die Gegenwart'. *Physikalische Blätter*, **24**: 530–7.

Hellman H (1998), 'Wegener versus everybody'. Chapter 8 in H Hellman, *Great Feuds in Science: Ten of the Liveliest Disputes Ever* (New York: John Wiley): 141–58.

Henderson L (2018), 'Global versus local arguments for realism'. In J Saatsi (ed.), *The Routledge Handbook of Scientific Realism* (Abingdon: Routledge): 151–63.

Hendry R (2020), 'Trusting atoms'. In U Zilioli (ed.), *Atomism in Philosophy: A History from Antiquity to the Present* (London: Bloomsbury): 470–88.

Henehan MJ, Ridgwell A, Thomas E, Zhang S, Alegret L, Schmidt DN, Rae JWB, Witts JD, Landman NH, Greene SE, Huber BT, Super J, Planavsky NJ, and Hull PM (2019), 'Rapid ocean acidification and phased biogeochemical recovery following the end-Cretaceous Chicxulub impact'. *Proceedings of the National Academy of Sciences* **116**(45): 22500–4.

Herzenberg CL (2008), 'Grete Hermann: an early contributor to quantum theory'. Arxiv 0812.3986v1, physics.gen-ph: 1–22. Available at https://arxiv.org/abs/0812.3986.

Hesse M (1976), 'Truth and the growth of scientific knowledge'. *PSA: Proceedings of the Biennial Meeting of the Philosophy of Science Association* **2**: 261–80.

Hoefer C (2020), 'Scientific realism without the quantum'. Chapter 2 in S French and J Saatsi (eds), *Scientific Realism and the Quantum* (Oxford: Oxford University Press).

Hoefer C and Martí G (2020), 'Realism, reference & perspective'. *European Journal for Philosophy of Science* **10**(3): article 38.

Holmes A (1928), 'Continental drift'. *Nature* **122**: 431–3.

Horsley S (1770), 'Difficulties in the Newtonian theory of light, considered and removed'. *Phil. Trans.* **60**: 417–40.

Howard J and Vogel J (2019), *Encyclopedia of Animals* (London: Quarto Publishing).

Hoyningen-Huene P (2022), 'Is Kuhn's "world change through revolutions" comprehensible?' Epistemology and Philosophy of Science, forthcoming 2022.

Hu B, Margolin W, Molineux IJ, and Liu J (2013), 'The bacteriophage T7 virion undergoes extensive structural remodeling during infection'. *Science* **339**(6119): 576–9.

Huang C, Wang Y, Li X, Ren L, Zhao J, Hu Y, Zhang L, Fan G, Xu J, Gu X, Cheng Z, Yu T, Xia J, Wei Y, Wu W, Xie X, Yin W, Li H, Liu M, Xiao Y, Gao H, Guo L, Xie J, Wang G, Jiang R, Gao Z, Jin Q, Wang J, and Cao B (2020), 'Clinical features of patients infected with 2019 novel coronavirus in Wuhan, China'. *The Lancet* **395**(10223): 497–506.

Huang J, Gates AJ, Sinatra R, and Barabási A (2020), 'Historical comparison of gender inequality in scientific careers across countries and disciplines'. *PNAS* **117**(9): 4609–16.

Hudson RG (1999), 'Mesosomes: a study in the nature of experimental reasoning'. *Philosophy of Science* **66**(2): 289–309.

Hudson RG (2013), *Seeing Things: The Philosophy of Reliable Observation* (New York: Oxford University Press).

Hull D (1988), *Science As a Process* (Chicago, IL: University of Chicago Press).

Hull PM, Bornemann A, Penman DE, Henehan MJ, Norris RD, Wilson PA, Blum P, Alegret L, Batenburg SJ, Bown PR, Bralower TJ, Cournede C, Deutsch A, Donner B, Friedrich O, Jehle S, Kim H, Kroon D, Lippert PC, Loroch D, Moebius I, Moriya K, Peppe DJ, Ravizza GE, Röhl U, Schueth JD, Sepúlveda J, Sexton PF, Sibert EC, Śliwińska KK, Summons RE, Thomas E, Westerhold T, Whiteside JH, Yamaguchi T, and Zachos JC (2020), 'On impact and volcanism across the Cretaceous-Paleogene boundary'. *Science* **367**(6475): 266–72.

Hutchinson J (1946), *A Botanist in Southern Africa* (London: PR Hawthorn).

Institute of Medicine (2008), *Science, Evolution, and Creationism* (Washington, DC: The National Academies Press); available at https://doi.org/10.17226/11876.

Intemann K (2017), 'Who Needs Consensus Anyway? Addressing Manufactured Doubt And Increasing Public Trust In Climate Science', *Public Affairs Quarterly* **31**(3): 189–208.

Irving E (1956), 'Palaeomagnetic and palaeoclimatological aspects of polar wandering'. *Geofis. Pura Appl.* **33**: 23–48.

Irving JCE (2018), 'Earth's soft heart: a modern seismological method raises questions about the properties of Earth's inner core'. *Science* **362**(6412): 294.

Ivanova M (2013), 'Did Perrin's experiments convert Poincaré to scientific realism?'. *The Journal of the International Society for the History of Philosophy of Science* **3**(1): 1–19.

Jagoda M and Rutkowska M (2020), 'Use of VLBI measurement technique for determination of motion parameters of the tectonic plates'. *Metrol. Meas. Syst.* **27**(1): 151–65.

Jaki SL (1969), *The Paradox of Olbers' Paradox* (New York: Herder and Herder).

Janzwood S (2020), 'Confident, likely, or both? The implementation of the uncertainty language framework in IPCC special reports'. *Climate Change* **162**: 1655–75.

Joleaud L (1924), 'L'histoire biogeographique de l'Amerique et la theorie de Wegener'. *J. Soc. Am. Paris* **16**: 325–60.

Jones RW (2006), *Applied Palaeontology* (Cambridge: Cambridge University Press).

Jongmans WJ (1937), 'The flora of the Upper Carboniferous of Djambi (Sumatra, Netherl., India) and its possible bearing on the paleogeography of the Carboniferous'. Compte Rendu, 2e Congress de Stratigraphie Carbonifere, Heerlen, 1935, **1**: 345–62.

Jordan JS (2013), 'Consciousness and embodiment'. In H Pashler (ed.), *Encyclopedia of the Mind* (Newbury Park, CA: SAGE Publications): 194–9.

Kang SK and Kaplan S (2019), 'Working toward gender diversity and inclusion in medicine: myths and solutions'. *Lancet* **393**(10171): 579–86.

Karaca K (2017), 'A case study in experimental exploration: exploratory data selection at the Large Hadron Collider', *Synthese* **194**(2): 333–54.

Keller G, Adatte T, Bajpai S, Mohabey DM, Widdowson M, Khosla A, Sharma R, Khosla SC, Gertsch B, Fleitmann D, and Sahni A (2009), 'K–T transition in Deccan Traps of central India marks major marine Seaway across India'. *Earth and Planetary Science Letters* **282**: 10–23.

Keller G, Adatte T, Pardos A, Bajpai U, Khosla A, and Samant B (2010), 'Cretaceous extinctions: evidence overlooked'. *Science* **328**: 974–5.

Keller G, Adatte T, Stinnesbeck W, Rebolledo-Vieyra M, Fucugauchi JU, Kramar U, and Stüben D (2004), 'Chicxulub impact predates the K-T boundary mass extinction'. *PNAS* **101**(11): 3753–8.

Keller G, Bhowmick PK, Upadhyay H, Dave A, Reddy AN, Jaiprakash BC, and Adatte T (2011), 'Deccan volcanism linked to the Cretaceous-Tertiary boundary mass extinction: new evidence from ONGC wells in the Krishna-Godavari Basin'. *Journal of the Geological Society of India* **78**: 399–428.

Keller G, Khozyem H, Adatte T, Malarkodi N, Spangenberg JE, and Stinnesbeck W (2013), 'Chicxulub impact spherules in the North Atlantic and Caribbean: age constraints and Cretaceous–Tertiary boundary hiatus'. *Geological Magazine* **150**(5): 885–907.

Keller G, Mateo P, Monkenbusch J, Thibault N, Punekar J, Spangenberg JE, Abramovich S, Ashckenazi-Polivoda S, Schoene B, Eddy MP, Samperton KM, Khadri SFR, and Adatte T (2020), 'Mercury linked to Deccan Traps volcanism, climate change and the end-Cretaceous mass extinction'. *Global and Planetary Change* **194**: 103312.

Keller G, Sahni A, and Bajpai S (2009), 'Deccan volcanism, the KT mass extinction and dinosaurs'. *Journal of Biosciences* **34**: 709–28.

Kew MC and Kassianides C (1996), 'HGV: hepatitis G virus or harmless G virus?'. *The Lancet* **348** (supplement II): 10.

Kidd IJ (2020), 'Humility, contingency, and pluralism in the sciences'. Chapter 30 in M Alfano, MP Lynch, and A Tanesini (eds), *The Routledge Handbook of Philosophy of Humility* (London: Routledge).

Kierulf HP (2017), 'Analysis strategies for combining continuous and episodic GNSS for studies of neo-tectonics in Northern Norway'. *Journal of Geodynamics* **109**: 32–40.

King L (1953), 'Necessity for continental drift'. *Bull. Am. Assoc. Petrol. Geol.* **37**: 2163–77.

Kitcher P (1993), *The Advancement of Science: Science Without Legend, Objectivity Without Illusions* (Oxford: Oxford University Press).

Kitcher P (2002), 'Scientific knowledge'. In P Moser (ed.), *The Oxford Handbook of Epistemology* (New York: Oxford University Press): 385–407.

Kitcher P (2019), 'What makes science trustworthy? Review of N Oreskes (2019), *Why Trust Science?* (Princeton, NJ: Princeton University Press)'. *Boston Review* (7 November 2019); available at https://bostonreview.net/articles/philip-kitcher-what-makes-science-trustworthy/.

Kragh H (1985), 'The fine structure of hydrogen and the gross structure of the physics community, 1916–26'. *Historical Studies in the Physical Sciences* **15**: 67–125.

Krige LJ (1926), 'On mountain building and continental sliding'. *S. Afr. J. Sci.* **23**: 206–15.

Krishnan MS (1959), '*History of the Indian Ocean*'. In M Sears (ed.), *International Oceanographic Congress* (Washington, DC: American Association for the Advancement of Science): 34–5.

Kronig R (1960), 'The turning point'. In M Fierz and VF Weisskopf (eds), *Theoretical Physics in the Twentieth Century* (New York: Interscience Publishers): 5–38.

Kuhn TS (1962), *The Structure of Scientific Revolutions*, 1st edition (Chicago, IL: University of Chicago Press).

Kuhn TS (1970), *The Structure of Scientific Revolutions*, 2nd edition (Chicago, IL: University of Chicago Press).

Kuhn TS (2012), *The Structure of Scientific Revolutions: 50th Anniversary Edition* (Chicago, IL: University of Chicago Press).

Ladyman J (2011), 'Structural realism versus standard scientific realism: the case of phlogiston and dephlogisticated air'. *Synthese* **180**: 87–101.
Lalchhandama K (2020), 'The chronicles of coronaviruses: the bronchitis, the hepatitis and the common cold'. *Science Vision* **20**(1): 43–53.
Landes J (2021), 'The variety of evidence thesis and its independence of degrees of independence'. *Synthese* **198**: 10611–41.
Landhuis E (2016), 'Scientific literature: information overload'. *Nature* **535**: 457–8.
Landsman NP (2006), 'When champions meet: rethinking the Bohr–Einstein debate'. *Studies in History and Philosophy of Modern Physics* **37**: 212–42.
Laudan L (1981), 'A confutation of convergent realism'. *Philosophy of Science* **48**, 19–48.
Laudan L (1984), *Science and Values: The Aims of Science and their Role in Scientific Debate* (Oakland, CA: University of California Press).
Lenoir T (1989), *The Strategy of Life: Teleology and Mechanics in Nineteenth-Century German Biology* (Chicago, IL: University of Chicago Press).
Letokhov VS and Johansson S (2009), *Astrophysical Lasers* (Oxford: Oxford University Press).
Lipton P (2001), 'Quests of a realist'. Part of a review symposium of S Psillos (1999), *Scientific Realism: How Science Tracks Truth* (Oxford: Oxford Univeresity Press). *Metascience* **10**: 347–53.
Longair M (2013), *Quantum Concepts in Physics: An Alternative Approach to the Understanding of Quantum Mechanics* (Cambridge: Cambridge University Press).
Longino H (1990), *Science as Social Knowledge: Values and Objectivity in Scientific Inquiry* (Princeton, NJ: Princeton University Press).
Longino H (2002), *The Fate of Knowledge* (Princeton, NJ: Princeton University Press).
Lowery CM, Bralower TJ, Owens JD, Rodríguez-Tovar FJ, Jones H, Smit J, Whalen MT, Claeys P, Farley K, Gulick SPS, Morgan JV, Green S, Chenot E, Christeson GL, Cockell CS, Coolen MJL, Ferrière L, Gebhardt C, Goto K, Kring DA, Lofi J, Ocampo-Torres R, Perez-Cruz L, Pickersgill AE, Poelchau MH, Rae ASP, Rasmussen C, Rebolledo-Vieyra M, Riller U, Sato H, Tikoo SM, Tomioka N, Urrutia-Fucugauchi J, Vellekoop J, Wittmann A, Xiao L, Yamaguchi KE, and Zylberman W (2018), 'Rapid recovery of life at ground zero of the end-Cretaceous mass extinction'. *Nature* **558**, 288–91.
Lu Y, Yoo B, Ng AHC, Kim J, Yeom S, Tang J, Lin MM, Zewail AH, and Heath JR (2019), '4D electron microscopy of T cell activation'. *PNAS* **116**(44): 22014–19.
Luo Y (2007), *Comprehensive Handbook of Chemical Bond Energies* (Boca Raton, FL: CRC Press).
Lyons SL, Karp AT, Bralower TJ, Grice K, Schaefer B, Gulick SPS, Morgan JV, and Freeman KH (2020), 'Organic matter from the Chicxulub crater exacerbated the K-Pg impact winter'. *PNAS* **117**(41): 25327–34.
Lyons TD (2006), 'Scientific realism and the stratagema de divide et impera'. *British Journal for the Philosophy of Science* **57**: 537–60.
Lyons TD and Vickers P (eds) (2021), *Contemporary Scientific Realism: The Challenge from the History of Science* (New York: Oxford University Press).
Maack R (1934), 'Die Godwanaschichten in Suedbrasilien und ihre Beziehungen zur Kaokoformation Suedwestafrikas'. *Z. Ges. Kdrdlk. Erdkunde Berlin* **5/6**: 194–222.
Mabe M and Amin M (2001), 'Growth dynamics of scholarly and scientific journals'. *Scientometrics* **51**(1): 147–62.
MacKinnon E (1982), *Scientific Explanation and Atomic Physics* (Chicago, IL: University of Chicago Press).

McMullin E (1984), 'A case for scientific realism'. In J Leplin (ed.), *Scientific Realism* (Berkeley, CA: University of California Press): 8–40.

Maddy P (2001), 'Naturalism: friends and foes'. *Philosophical Perspectives* **15**: 37–67.

Magnus PD and Callender C (2004), 'Realist ennui and the base rate fallacy'. *Philosophy of Science* **71**(3): 320–38.

Marini M, Falqui A, Moretti M, Limongi T, Allione M, Genovese A, Lopatin S, Tirinato L, Das G, Torre B, Giugni A, Gentile F, Candeloro P, and Fabrizio ED (2015), 'The structure of DNA by direct imaging'. *Science Advances* **1**(7); http://doi.org/10.1126/sciadv.1500734.

Martin H (1961), 'The hypothesis of continental drift in the light of recent advances of geological knowledge in Brazil and in South West Africa: Alex. L. du Toit memorial lectures No. 7'. *Trans. Proc. Geol. Soc. S. Afr.* (Annexure to), **64**: 1–47.

Mayer-Scholl A, Hurwitz R, Brinkmann V, Schmid M, Jungblut P, Weinrauch Y, and Zychlinsky A (2005), 'Human neutrophils kill *Bacillus anthracis*'. *PLoS Pathogens* **1**(3): e23.

Mayo DG (1991), 'Novel evidence and severe tests'. *Philosophy of Science* **58**(4): 523–52.

Meckel JF (1811), *Entwurf einer Darstellung der zwischen dem Embryozustande der höheren Tiere und dem permanenten der niederen stattfindenen Parallele: Beyträge zur vergleichenden Anatomie*, Volume 2 (Leipzig: Carl Heinrich Reclam): 1–60.

Meehan C and Curry RL (2008), 'Exploitation by a predominantly vegetarian jumping spider (*Bagheera kiplingi*) of Acacia mutualism through "'clever" avoidance of ant-guards'. 12th International Behavioral Ecology Congress, 9–15 August 2008 (Ithaca, NY).

Menz H (2000), *Martin Heinrich Rathke (1793–1860): ein Embryologe des 19. Jahrhunderts.* Acta biohistorica (Marburg: Basilisken-Presse).

Mercier H and Sperber D (2017), *The Enigma of Reason* (Cambridge, MA: Harvard University Press).

Mermin ND (1993), 'Hidden variables and the two theorems of John Bell'. *Reviews of Modern Physics* **65**: 803–15.

Michelson AA (1903), *Light Waves and Their Uses* (Chicago, IL: University of Chicago Press).

Miller B (2013), 'When is consensus knowledge based? Distinguishing shared knowledge from mere agreement'. *Synthese* **190**(7): 1293–316.

Miller B (2019), 'The social epistemology of consensus and dissent'. Chapter 22 in M Fricker, PJ Graham, D Henderson, and NJLL Pedersen (eds), *The Routledge Handbook of Social Epistemology*, 1st edition (London: Routledge): 230–9.

Milliken A (2017), 'Subjective from the start: a critique of transformative criticism'. *Nursing Philosophy* **18**(2): e12156.

Miyake T and Smith G (2021), 'Realism, physical meaningfulness, and molecular spectroscopy'. Chapter 8 in TD Lyons and P Vickers (eds), *Contemporary Scientific Realism: The Challenge from the History of Science* (New York: Oxford University Press).

Moody PA (1953), *Introduction to Evolution* (New York: Harper & Bros).

Moss-Racusin CA, van der Toorn J, Dovidio JF, Brescoll VL, Graham MJ, and Handelsman J (2016), 'A "scientific diversity" intervention to reduce gender bias in a sample of life scientists'. *CBE Life Sci Educ.* **15**(3): ar29.

Musgrave A (1974), 'Logical versus historical theories of confirmation'. *British Journal for the Philosophy of Science* **25**: 1–23.

Musgrave A (1976), 'Why did oxygen supplant phlogiston? Research programmes in the chemical revolution'. In C Howson (ed.), *Method and Appraisal in the Physical Sciences* (Cambridge: Cambridge University Press): 181–209.

Musgrave A (1985), 'Realism versus constructive empiricism'. In PM Churchland and CA Hooker (eds), *Images of Science* (Chicago, IL: Chicago University Press): 197–221.

Nanninga N (1971), 'The mesosome of *Bacillus subtilis* as affected by chemical and physical fixation'. *Journal of Cell Biology* **48**(1): 219–24.

Nanninga N (1973), 'Freeze-fracturing of microorganisms: physical and chemical fixation of *Bacillus subtilis*'. In E Benedetti and P Favard (eds), *Freeze-Etching: Techniques and Applications* (Paris: Socittir Francaise de Microscopic Electronique): 151–79.

Nascimento-Silva MV, Sial AN, Ferreira VP, Neumann VH, Barbosa JA, Pimentel MM, and de Lacerda LD (2011), 'Cretaceous-Paleogene transition at the Paraíba Basin, northeastern Brazil: carbon-isotope and mercury subsurface stratigraphies'. *Journal of South American Earth Sciences* **32**(4): 379–92.

Neuman BW, Adair BD, Yoshioka C, Quispe JD, Orca G, Kuhn P, Milligan RA, Yeager M, and Buchmeier MJ (2006), 'Supramolecular architecture of severe acute respiratory syndrome coronavirus revealed by electron cryomicroscopy'. *Journal of Virology* **80** (16): 7918–28.

Nicholls JM, Butany J, Poon LL, Chan KH, Beh SL, Poutanen S, Peiris JS, and Wong M (2006), 'Time course and cellular localization of SARS-CoV nucleoprotein and RNA in lungs from fatal cases of SARS'. *PLoS Medicine* **3**(2): e27.

Niedzwiedzki G, Szrek P, Narkiewicz K, Narkiewicz M, and Ahlberg PE (2010), 'Tetrapod trackways from the early Middle Devonian period of Poland'. *Nature* **463**: 43–8.

Niiniluoto I (2020), 'Social aspects of scientific knowledge'. *Synthese* **197**: 447–68.

Nisio S (1973), 'The formation of the Sommerfeld quantum theory of 1916'. *Japanese Studies in the History of Science* **12**: 39–78.

Nitecki MH, Lemke JL, Pullman HW, and Johnson ME (1978), 'Acceptance of plate tectonic theory by geologists'. *Geology* **6**(11): 661–4.

Norton JD (2021), *The Material Theory of Induction* (Calgary: University of Calgary Press).

Noyes RW (1982), *The Sun: Our Star* (Cambridge, MA: Harvard University Press).

O'Connor T and Robb D (eds) (2003), *Philosophy of Mind: Contemporary Readings* (London: Routledge).

Oliveira AI and Leonardos OH (1940), *Geologia do Brasil* (Rio de Janeiro: n.p.) [2nd edition, 1943].

Oreskes N (2004), 'The scientific consensus on climate change'. *Science* **306**(5702): 1686.

Oreskes N (2019), *Why Trust Science?* (Princeton, NJ: Princeton University Press).

Oreskes N and Conway EM (2010), *Merchants of Doubt* (New York: Bloomsbury).

Ospovat D (1976), 'The influence of Karl Ernst von Baer's embryology, 1828–1859: a reappraisal in light of Richard Owen's and William B. Carpenter's "Palaeontological Application of Von Baer's Law"'. *Journal of the History of Biology* **9**(1): 1–28.

Oteyza DG de, Gorman P, Chen Y, Wickenburg S, Riss A, Mowbray DJ, Etkin G, Pedramrazi Z, Tsai H, Rubio A, Crommie MF, and Fischer FR (2013), 'Direct imaging of covalent bond structure in single-molecule chemical reactions'. *Science* **340**(6139): 1434–7.

Pais A (1991), *Niels Bohr's Times* (Oxford: Oxford University Press).

Palmer E (1954), 'Investigations of the gastric mucosa spirochetes of the human'. *Gastroenterology* **27**: 218–20.

Parker W (2020), 'Evidence and knowledge from computer simulation'. *Erkenntnis*; available at https://doi.org/10.1007/s10670-020-00260-1.

Pashby T (2012), 'Dirac's prediction of the positron: a case study for the current scientific realism debate'. *Perspectives on Science* **20**(4): 440–75.

Patch C and Landman O (1971), 'Comparison of the biochemistry and rates of synthesis of mesosomal and peripheral membranes in *Bacillus subtilis*'. *Journal of Bacteriology* **107**: 345–57.

Pearlman M, Arnold D, Davis M, Barlier F, Biancale R, Vasiliev V, Ciufolini I, Paolozzi A, Pavlis EC, Sośnica K, and Bloßfeld M (2019), 'Laser geodetic satellites: a high-accuracy scientific tool'. *Journal of Geodesy* **93**: 2181–94.

Petersen SV, Dutton A, and Lohmann KC (2016), 'End-Cretaceous extinction in Antarctica linked to both Deccan volcanism and meteorite impact via climate change'. *Nature Communications* **7**: 12079.

Playfair J (1805), 'Biographical account of the late Dr. James Hutton'. *Earth and Environmental Science Transactions of the Royal Society of Edinburgh* **5**: 39–99.

Plumstead E (1961), 'Ancient plants and drifting continents'. *S. Afr. J. Sci.* **57**: 173–81.

Pope K, Baines K, Ocampo A, and Ivanov B (1997), 'Energy, volatile production, and climatic effects of the Chicxulub Cretaceous/Tertiary impact'. *Journal of Geophysical Research* **102**(E9): 21645–64.

Popper K (1962), *Conjectures and Refutations* (New York: Basic Books).

Popper K (1970), 'Normal science and its dangers'. In I Lakatos and A Musgrave (eds), *Criticism and the Growth of Knowledge* (Cambridge: Cambridge University Press): 51–8.

Prialnik D (2009), *An Introduction to the Theory of Stellar Structure and Evolution* (Cambridge: Cambridge University Press).

Psillos S (1999), *Scientific Realism: How Science Tracks Truth* (Oxford: Oxford University Press).

Pyle A (2000), 'The rationality of the chemical revolution'. In R Nola and H Sankey (eds), *After Popper, Kuhn and Feyerabend* (Dordrecht: Kluwer): 99–124.

Rasmussen N (1993), 'Facts, artifacts, and mesosomes: practicing epistemology with a microscope'. *Studies in History and Philosophy of Science* **24**: 227–65.

Rasmussen N (2001), 'Evolving scientific epistemologies and the artifacts of empirical philosophy of science: a reply concerning mesosomes'. *Biology and Philosophy* **16**: 627–52.

Rathke H (1825a), 'Kiemen bei Säugthieren' ['Gills in mammals']. Oken's *Isis*: col. 747–9.

Rathke H (1825b), 'Kiemen bei Vögeln' ['Gills in birds']. Oken's *Isis*: col. 1100–1.

Rathke H (1825c), 'Beobachtungen und Betrachtungen über die Entwicklung der Geschlechtswerkzeuge bei den Wirbelthieren'. *Neueste Schriften. naturforsch. Gesellsch., Danzig* **1** (Heft 4): 1–146.

Reagan RL, Hauser JE, Lillie MG, Craige Jr AH (1948), 'Electron micrograph of the virus of infectious bronchitis of chickens'. *The Cornell Veterinarian* **38**(2): 190–1.

Reich ES (2012), 'Embattled neutrino project leaders step down'. *Nature* (2 April 2012); https://doi.org/10.1038/nature.2012.10371.

Renne PR, Deino AL, Hilgen FJ, Kuiper KF, Mark DF, Mitchell WS 3rd, Morgan LE, Mundil R, and Smit J (2013), 'Time scales of critical events around the Cretaceous-Paleogene boundary'. *Science* **339**(6120): 684–7.

Renne PR, Sprain CJ, Richards MA, Self S, Vanderkluysen L, and Pande K (2015): 'State shift in Deccan volcanism at the Cretaceous-Paleogene boundary, possibly induced by impact'. *Science* **350**(6256): 76–8.

Rogers EM (2003), *Diffusion of Innovations*, 5th edition (New York: Simon and Schuster).

Rogers HJ (1970), 'Bacterial growth and the cell envelope'. *Bacteriological Reviews* **34**(2): 194–214.

Rossetter T (2018), 'Realism on the rocks: novel success and James Hutton's theory of the earth'. *Studies in History and Philosophy of Science* **67**: 1–13.

Rossetter T (2022), 'From crisis, to normal science, and back again: coming full Kuhn cycle in the career of Warren B. Hamilton'. In GR Foulger, DM Jurdy, CA Stein, LC Hamilton, K Howard, and S Stein (eds), *In the Footsteps of Warren B. Hamilton: New Ideas in Earth*

Science: Geological Society of America Special Paper 553 (Boulder, CO: Geological Society of America), pp. 7–19.

Rowbottom D (2019), *The Instrument of Science: Scientific Anti-Realism Revitalised* (London: Routledge).

Russell ES (1916), *Form and Function* (London: J. Murray).

Ryter A (1969), 'Structure and functions of mesosomes of gram positive bacteria'. *Current Topics in Microbiology and Immunology* **49**: 151–77.

Saatsi J (2009), 'Grasping at realist straws'. Review symposium of Stanford (2006), *Metascience* **18**: 355–63.

Saatsi J (2017), 'Replacing recipe realism'. *Synthese* **194**: 3233–44.

Saatsi J (ed.) (2018), *The Routledge Handbook of Scientific Realism* (London: Routledge).

Saatsi J (2020), 'Truth vs. progress realism about spin'. Chapter 3 in S French and J Saatsi (eds), *Scientific Realism and the Quantum* (Oxford: Oxford University Press).

Saatsi J and Vickers P (2011), 'Miraculous success? Inconsistency and untruth in Kirchhoff's diffraction theory'. *British Journal for the Philosophy of Science* **62**(1): 29–46.

Sahni B (1936), 'Wegener's theory of continental drift in the light of palaeobotanical evidence'. *Indian Bot. Soc. J.* **15**: 319–32.

Schindler S (2018), *Theoretical Virtues in Science: Uncovering Reality through Theory* (Cambridge: Cambridge University Press).

Schoene B, Eddy MP, Samperton KM, Keller CB, Keller G, Adatte T, and Khadri SFR (2019), 'U-Pb constraints on pulsed eruption of the Deccan Traps across the end-Cretaceous mass extinction'. *Science* **363**(6429): 862–6.

Schoene B, Samperton KM, Eddy MP, Keller G, Adatte T, Bowring SA, Khadri SFR, and Gertsch B (2015), 'U-Pb geochronology of the Deccan Traps and relation to the end-Cretaceous mass extinction'. *Science* **347**(6218): 182–4.

Schulte P, Alegret L, Arenillas I, Arz JA, Barton PJ, Bown PR, Bralower TJ, Christeson GL, Claeys P, Cockell CS, Collins GS, Deutsch A, Goldin TJ, Goto K, Grajales-Nishimura JM, Grieve RAF, Gulick SPS, Johnson KR, Kiessling W, Koeberl C, Kring DA, MacLeod KG, Matsui T, Melosh J, Montanari A, Morgan JV, Neal CR, Nichols DJ, Norris RD, Pierazzo E, Ravizza G, Rebolledo-Vieyra M, Reimold WU, Robin E, Salge T, Speijer RP, Sweet AR, Urrutia-Fucugauchi J, Vajda V, Whalen MT, and Willumsen PS (2010), 'The Chicxulub asteroid impact and mass extinction at the Cretaceous-Paleogene boundary'. *Science* **327** (5970): 1214–18.

Scott EC and Branch G (2009), 'Don't call it "Darwinism"'. *Evolution: Education and Outreach* **2**: 90–4.

Scriven M (1959), 'Explanation and prediction in evolutionary theory', *Science* **130**(3374): 477–82.

Seevinck M (2017), 'Challenging the gospel: Grete Hermann on von Neumann's no-hidden-variables proof'. Chapter 7 in E Crull and G Bacciagaluppi (eds), *Grete Hermann—Between Physics and Philosophy* (Dordrecht: Springer): 107–18.

Šešelja D and Straßer C (2014), 'Heuristic reevaluation of the bacterial hypothesis of peptic ulcer disease in the 1950s'. *Acta Biotheor* **62**: 429–54.

Seward A and Conway V (1934), 'A phytogeographical problem: fossil plants from the Kerguelen Archipelago'. *Ann. Bot.* **48**: 736–7.

Sharp Z (2009), *Annual Editions: Environment: 09/10* (New York: McGraw-Hill).

Shayeghi A, Rieser P, Richter G, Sezer U, Rodewald JH, Geyer P, Martinez TJ, and Arndt M (2020), 'Matter-wave interference of a native polypeptide'. *Nature Communications* **11**: article 1447.

Shubin NH, Daeschler EB, and Jenkins Jr FA (2006), 'The pectoral fin of *Tiktaalik roseae* and the origin of the tetrapod limb'. *Nature* **440**: 764–71.

Shwed U and Bearman PS (2010), 'The temporal structure of scientific consensus formation'. *American Sociological Review* **75**(6): 817–40.

Sial AN, Chen J, Lacerda LD, Peralta S, Gaucher C, Frei R, Cirilli S, Ferreira VP, Marquillas RA, Barbosa JA, Pereira NS, and Belmino IKC (2014), 'High-resolution Hg chemostratigraphy: a contribution to the distinction of chemical fingerprints of the Deccan volcanism and Cretaceous–Paleogene boundary impact event'. *Palaeogeography, Palaeoclimatology, Palaeoecology* **414**: 98–115.

Silva MT (1966a), 'Novos dados sobre a fixaçao das estructuras membranosas das bacterias. Observaçoes em Bacillus cereus e Nocardia asteroides' ['New data on the attachment of bacterial membrane structures. Observations on *Bacillus cereus* and *Nocardia asteroides*']. Reunião Anual da Sociedade Portuguesa de Microscopia Electrónica (SPME): Resumos das Comunicações (Proceedings of the I Annual Meeting of the Portuguese Society for Electron Microscopy): Abstract 11.

Silva MT (1966b), 'Studies on the fixation of the mesosomes of some gram positive bacteria for electron microscopy'. *Sixth International Congress for Electron Microscopy*, Kyoto, Vol. 2 (Nihonbashi, Tokyo: Maruzen Co.): 275.

Silva MT (1967a), 'Influencia da concentracao de OsO_4, Cl_2Ca e tampão de acetato-veronal na fixacao dos mesosomas de "Bacillus cereus"' ['Influence of the concentration of OsO_4, Cl_2Ca and acetate-veronal buffer on the fixation of "Bacillus cereus" mesosomes']. Reunião Anual da Sociedade Portuguesa de Microscopia Electrónica (SPME): Resumos das Comunicações (Proceedings of the II Annual Meeting of the Portuguese Society for Electron Microscopy): Abstract 26.

Silva MT (1967b), 'Ultrastrutura do "Bacillus anthracis" in vivo' ['Ultrastructure of *Bacillus anthracis* in vivo']. Reunião Anual da Sociedade Portuguesa de Microscopia Electrónica (SPME): Resumos das Comunicações (Proceedings of the II Annual Meeting of the Portuguese Society for Electron Microscopy): Abstract 28.

Silva MT (1971), 'Changes induced in the ultrastructure of the cytoplasmic and intracytoplasmic membranes of several gram-positive bacteria by variations in OsO_4 fixation'. *Journal of Microscopy* **93**(3): 227–32.

Silva MT and Abram D (1971), 'Modificações na ultrastructura das membranas de *Nocardia asteroides* induzidas por variações nas técnicas de fixação quimica e coloração negativa' ['Modifications in the ultrastructure of *Nocardia asteroides* membranes induced by variations in chemical fixation and negative staining techniques']. Reunião Anual da Sociedade Portuguesa de Microscopia Electrónica (SPME): Resumos das Comunicações (Proceedings of the VI Annual Meeting of the Portuguese Society for Electron Microscopy): Abstract 12.

Silva MT and Da Fonseca AF (1970), 'Ultrastructure comparison between *Sarcina (Sporosarcina) ureae* and *Bacillus pasteurii*'. Reunião Anual da Sociedade Portuguesa de Microscopia Electrónica (SPME): Resumos das Comunicações (Proceedings of the V Annual Meeting of the Portuguese Society for Electron Microscopy): Abstract 33.

Silva MT and Guerra FC (1967), 'Accao de acetato de uranilo na preservacao das membranas em microscopia electronica. Estudo estrutural e bioquimico em "Bacillus subtilis" e cerebro de Rato' ['Action of uranyl acetate in the preservation of membranes in electron microscopy. Structural and biochemical study in "Bacillus subtilis" and rat's brain']. Reunião Anual da Sociedade Portuguesa de Microscopia Electrónica (SPME): Resumos das Comunicações (Proceedings of the II Annual Meeting of the Portuguese Society for Electron Microscopy): Abstract 27.

Silva MT, Paula Lima M, and Castro Melo JV (1969), 'Ultraestrutura da Nocardia coeliaca' ['Ultrastructure of *Nocardia coeliaca*']. Reunião Anual da Sociedade Portuguesa de

Microscopia Electrónica (SPME): Resumos das Comunicações (Proceedings of the IV Annual Meeting of the Portuguese Society for Electron Microscopy): Abstract 6.

Silva MT and Santos Mota JM (1968), 'Sobre a afinidade do acetato de uranilo para as membranas de algumas Bacterias' ['On the affinity of uranyl acetate for the membranes of some Bacteria']. Reunião Anual da Sociedade Portuguesa de Microscopia Electrónica (SPME): Resumos das Comunicações (Proceedings of the III Annual Meeting of the Portuguese Society for Electron Microscopy): Abstract 9.

Simpson GG (1943), 'Mammals and the nature of continents'. *Am. J. Sci.* **241**: 1–31.

Smart JJC (1968), *Between Science and Philosophy: An Introduction to the Philosophy of Science* (New York: Random House).

Solomon M (2001), *Social Empiricism* (Cambridge, MA: MIT Press).

Solomon M (2015), *Making Medical Knowledge* (Oxford: Oxford University Press).

Sömmerring ST von (1799), *Icones embryonum humanorum* (Frankfurt am Main: n.p.).

Sprenger J and Hartmann S (2019), *Bayesian Philosophy of Science* (New York: Oxford University Press).

Stanford PK (2006), *Exceeding Our Grasp: Science, History, and the Problem of Unconceived Alternatives* (Oxford: Oxford University Press).

Stanford PK (2009), 'Author's response'. In 'Grasping at realist straws', a review symposium of Stanford (2006). *Metascience* **18**: 379–90.

Stanford PK (2011), 'Damn the consequences: projective evidence and the heterogeneity of scientific confirmation'. *Philosophy of Science* **78**(5): 887–99.

Stanford PK (2015a), '"Atoms exist" is probably true, and other facts that should not comfort scientific realists'. *Journal of Philosophy* **112**(8): 397–416.

Stanford PK (2015b), 'Catastrophism, uniformitarianism, and a scientific realism debate that makes a difference'. *Philosophy of Science* **82**(5): 867–78.

Stanford PK (2019), 'Unconceived alternatives and conservatism in science: the impact of professionalization, peer-review, and Big Science'. *Synthese* **196**(10): 3915–32.

Stanford PK (2020), 'Resisting scientific realism with *or Without* van Fraassen's Darwinian explanation'. Review of Wray KB (2018), *Resisting Scientific Realism* (Cambridge: Cambridge University Press). *Metascience* **29**: 25–31.

Stanford PK (2021), 'Realism, instrumentalism, particularism: a middle path forward in the scientific realism debate'. Chapter 10 in TD Lyons and P Vickers (eds), *Contemporary Scientific Realism: The Challenge from the History of Science* (New York: Oxford University Press).

Steed P and Murray RG (1966), 'The cell wall and cell division of gram negative bacteria'. *Canadian Journal of Microbiology* **12**: 263–70.

Stegenga J (2016), 'Three criteria for consensus conferences'. *Foundations of Science* **21**(1): 35–49.

Stein H (1989), 'Yes, but... some skeptical remarks on realism and anti-realism'. *Dialectica* **43**(1/2): 47–65.

Steinle F (1997), 'Entering new fields: exploratory uses of experimentation'. *Philosophy of Science* **64** (Proceedings): S65–S74.

Steinle F (2002), 'Experiments in history and philosophy of science'. *Perspectives on Science* **10**: 408–32.

Stewart JA (1986), 'Drifting continents and colliding interests: a quantitative application of the interests perspective'. *Social Studies of Science* **16**(2): 261–79.

Strevens M (2003), 'The role of the priority rule in science'. *Journal of Philosophy* **100**(2): 55–79.

Suess FE (1937/8/9), 'Bausteine zu einem System der Tektogenese'. *Fortschr. Geol. Palaeontol.* **13**: 1–86; 87–238; 239–376.

Swisher 3rd CC, Grajales-Nishimura JM, Montanari A, Margolis SV, Claeys P, Alvarez W, Renne P, Cedillo-Pardoa E, Maurrasse FJ, Curtis GH, Smit J, and McWilliams MO (1992), 'Coeval $^{40}Ar/^{39}Ar$ ages of 65.0 million years ago from Chicxulub crater melt rock and Cretaceous-Tertiary boundary tektites'. *Science* **257**(5072): 954–8.

Tan W, Zhao X, Ma X, Wang W, Niu P, Xu W, Gao GF, and Wu G (2020), 'Notes from the field: a novel coronavirus genome identified in a cluster of pneumonia cases—Wuhan, China 2019–2020'. *China CDC Weekly* **2**: 61–2.

Taylor FB (1910), 'Bearing of the Tertiary mountain belt on the origin of the Earth's plan'. *Bull. Geol. Soc. Am.* **21**: 179–226.

Taylor H and Vickers P (2017), 'Conceptual fragmentation and the rise of eliminativism'. *European Journal for Philosophy of Science* **7**: 17–40.

Tiedemann F (1808), *Zoologie zu seinen Vorlesungen entworfen. 1. Teil: Zoologie der Menschen und der Säugetiere* (Landshut: n.p.).

Tollefsen DP (2006), 'Group deliberation, social cohesion, and scientific teamwork: is there room for dissent?'. *Episteme* **3**(1–2): 37–51.

Tucker A (2003), 'The epistemic significance of consensus'. *Inquiry* **46**(4): 501–21.

Valentin G (1835), *Handbuch der Entwicklungsgeschichte des Menschen mit vergleichender Rücksicht der Entwicklung der Säugetiere und Vögel* (Berlin: Rücker).

Van Fraassen B (1980), *The Scientific Image* (Oxford: Clarendon Press).

Van Fraassen B (2001), 'Constructive empiricism now'. *Philosophical Studies* **106**: 151–70.

Van Iterson W (1963), 'Membranous structures in microorganisms'. In *Recent Progress in Microbiology*, VIII International Congress of Microbiology, 1962 (Toronto: University of Toronto Press): 14–37.

Vellekoop J, Sluijs A, Smit J, Schouten S, Weijers JWH, Sinninghe Damsté JS, and Brinkhuis H (2014), 'Rapid short-term cooling following the Chicxulub impact at the Cretaceous-Paleogene boundary'. *PNAS* **111**(21): 7537–41.

Vickers P (2013a), 'A confrontation of convergent realism'. *Philosophy of Science* **80**(2): 189–211.

Vickers P (2013b), *Understanding Inconsistent Science* (Oxford: Oxford University Press).

Vickers P (2014), 'Scientific theory eliminativism'. *Erkenntnis* **79**(1): 111–26.

Vickers P (2017), 'Understanding the selective realist defence against the PMI'. *Synthese* **194**(9): 3221–32.

Vickers P (2018a), 'Historical challenges to realism'. Chapter 4 in J Saatsi (ed.), *The Routledge Handbook of Scientific Realism* (London: Routledge): 48–59.

Vickers P (2018b), 'The misleading evidence that fooled scientists for decades'. *The Conversation*, June 2018; available at https://theconversation.com/the-misleading-evidence-that-fooled-scientists-for-decades-95737.

Vickers P (2019), 'Towards a realistic success-to-truth inference for scientific realism'. *Synthese* **196**(2): 571–85.

Vickers P (2020a), 'Disarming the ultimate historical challenge to scientific realism'. *British Journal for the Philosophy of Science* **71**(3): 987–1012.

Vickers P (2020b), 'Resisting scientific anti-realism'. Review of Wray KB (2018), *Resisting Scientific Realism* (Cambridge: Cambridge University Press). *Metascience* **29**: 11–16.

Vijgen L, Moës E, Keyaerts E, Li S, and Van Ranst M (2008), 'A pancoronavirus RT-PCR assay for detection of all known coronaviruses'. In D Cavanagh (ed.), *SARS and Other Coronaviruses: Methods in Molecular Biology (Methods and Protocols)* **454** (Totowa, NJ: Humana Press).

Von Baer KE (1826), 'The relations of affinity among the lower forms of animals'. *Nova Acta Physico Medica*, t. xiii. Translated in 'Fragments relating to philosophical zoology. Selected from the works of K. E. von Baer', in A Henfrey and TH Huxley (eds), *Scientific*

Memoirs, Selected from the Transactions of Foreign Academies of Science, and from Foreign Journals of Natural History (London: Taylor and Francis, 1853): 176–238.

Von Baer KE (1827), 'Ueber die Kiemen und Kiemengefässe in den Embryonen der Wirbelthiere' ['On the gills and gill vessels in the embryos of vertebrates']. *Meckel's Archiv für Anatomie und Physiologie* **1827**(4): 556–68.

Von Baer KE (1828), 'Über Entwickelungsgeschichte der Thiere. Beobachtung und Reflexion'. Translated in 'Fragments relating to philosophical zoology. Selected from the works of K. E. von Baer', in A Henfrey and TH Huxley (eds), *Scientific Memoirs, Selected from the Transactions of Foreign Academies of Science, and from Foreign Journals of Natural History* (London: Taylor and Francis, 1853): 176–238.

von Neumann J (1932), *Mathematische Grundlagen der Quantenmechanik* [*Mathematical Foundations of Quantum Mechanics*] (Berlin: Springer).

Votsis I (2007), 'Review of Kyle Stanford's "Exceeding our Grasp: Science, History and the Problem of Unconceived Alternatives"'. *International Studies in the Philosophy of Science* **21**(1): 103–6.

Wade A (1941), 'The geology of the Antarctic Continent and its relationship to neighbouring land areas'. *Proc. Roy. Soc. Queensland* **52**: 24–35.

Wallace RA and Sanders GP (1997), *Biology: The World of Life* (San Francisco, CA: Benjamin/Cummings).

Walton J (1953), *An Introduction to the Study of Fossil Plants*, 2nd edition (London: Adam and Charles Black).

Wang W, Xu Y, Gao R, Lu R, Han K, Wu G, and Tan W (2020), 'Detection of SARS-CoV-2 in different types of clinical specimens'. Research Letter, *Journal of the American Medical Association (JAMA)* **323**(18): 1843–4.

Washington H (2013), *Climate Change Denial: Heads in the Sand* (London: Routledge).

Waters CK (2007), 'The nature and context of exploratory experimentation: an introduction to three case studies of exploratory research'. *History and Philosophy of the Life Sciences* **29**: 275–84.

Wegener A (1915), *Die Entstehung der Kontinente und Ozeane* (Braunschweig: Friedrich Vieweg & Sohn). 1st edition, 1915; 2nd edition, 1920, 3rd edition, 1922; 4th edition, 1924; 4th revised edition, 1929; 5th edition, revised by K Wegener (1936) [translated as *The Origin of Continents and Oceans* (London: Methuen, 1924)].

Wegener A (1929), *Die Entstehung der Kontinente und Ozeane*, 4th revised edition (Braunschweig: Friedrich Vieweg & Sohn).

Wells GA (1967a), 'Goethe and the intermaxillary bone'. *The British Journal for the History of Science* **3**(4): 348–61.

Wells GA (1967b), 'Goethe and evolution'. *Journal of the History of Ideas* **28**(4): 537–50.

Whewell W (1849), *Of Induction: With Especial Reference to Mr. J. Stuart Mill's System of Logic* (London: John W. Parker).

White NJ, Snook RR, and Eyres I (2020), 'The past and future of experimental speciation'. *Trends in Ecology and Evolution* **35**(1): 10–21.

Wiles JR (2010), 'Overwhelming Scientific Confidence in Evolution and Its Centrality in Science Education—And the Public Disconnect', *Science Education Review* **9**(1): 18–27.

Williams MB (1973), 'Falsifiable predictions of evolutionary theory'. *Philosophy of Science* **40**(4): 518–37.

Williams SA (2018), 'Was the last common ancestor aping a chimp or just monkeying around?'. *Journal of Human Evolution* **121**(August): 72–4.

Willis B (1944), 'Continental drift, ein märchen'. *Am. J. Sci.* **242**: 509–13.

Wilson M (2006), *Wandering Significance* (Oxford: Oxford University Press).

Windhausen A (1931), *Geología Argentina (Segunda Parte)* (Buenos Aires).
Worrall J (2007), 'Miracles and models: why reports of the death of structural realism may be exaggerated'. *Royal Institute of Philosophy Supplements* **61**: 125–54.
Wray KB (2018), *Resisting Scientific Realism* (Cambridge: Cambridge University Press).
Wray KB (2020), 'Author's response'. *Metascience* **29**: 33–40.
Wulff, EV (1950), *An Introduction to Historical Plant Geography*. Translated by E Brissenden (Waltham, MA: Chronica Botanica).
Wylie A (2006), 'Socially naturalized norms of epistemic rationality: aggregation and deliberation'. *The Southern Journal of Philosophy* **44**: 43–8.
Zaki AM, Van Boheemen S, Bestebroer TM, Osterhaus ADME, and Fouchier RAM (2012), 'Isolation of a novel coronavirus from a man with pneumonia in Saudi Arabia'. *New England Journal of Medicine* **367**: 1814–20.
Zinkernagel H (2016), 'Niels Bohr on the wave function and the classical/quantum divide'. *Studies in History and Philosophy of Modern Physics* **53**: 9–19.
Zollman K (2010), 'The epistemic benefit of transient diversity'. *Erkenntnis* **72**: 17–35.

Index

Note: Figures are indicated by an italic '*f*', respectively, following the page number.

For the benefit of digital users, indexed terms that span two pages (e.g., 52–53) may, on occasion, appear on only one of those pages.